JASON SHUTE is an acknowledged expert on immigrant communities in South Australia and on Henry Ayers. As a prize-winning graduate of London's Royal College of Music, he pursued a career as a baritone soloist and choral and orchestral conductor. He has lived with his family in South Australia for almost two decades, where he continued working as a conductor and composer before turning to historical research and writing.

'This biography breathes wonderful new life into a forgotten giant of colonial South Australia, Sir Henry Ayers: mining tycoon, leading politician, women's rights and education reformer, coloniser of the Northern Territory, and godfather of the overland telegraph which first tethered Australian cable communications to the wider world.'

*– Carl Bridge, Head of the Menzies Centre
for Australian Studies, King's College London.*

HENRY AYERS

THE MAN WHO BECAME A ROCK

Jason Shute

Published in 2011 by I.B.Tauris & Co Ltd
6 Salem Road, London W2 4BU
175 Fifth Avenue, New York NY 10010
www.ibtauris.com

Distributed in the United States and Canada Exclusively by Palgrave Macmillan
175 Fifth Avenue, New York NY 10010

Copyright © Jason Shute, 2011

The right of Jason Shute to be identified as the author of this work has been asserted by the author in accordance with the Copyright, Designs and Patent Act 1988.

All rights reserved. Except for brief quotations in a review, this book, or any part thereof, may not be reproduced, stored in or introduced into a retrieval system, or transmitted, in any form or by any means, electronic, mechanical, photocopying, recording or otherwise, without the prior written permission of the publisher.

ISBN: 978 1 84885 563 2

A full CIP record for this book is available from the British Library
A full CIP record is available from the Library of Congress

Library of Congress Catalog Card Number: available

Typeset in Janson Text by Pindar NZ, Auckland, New Zealand
Printed and bound in Great Britain by CPI Antony Rowe, Chippenham

This book is published with generous support from History SA

CONTENTS

Illustrations		vii
Map of the Province of South Australia		ix
Family Trees		x
Preface		xii
Acknowledgements		xiv
Introduction		xviii
1	'In search of a place to ascend'	1
2	The Secretary	14
3	Friends when and where you need them	27
4	The Year of Evolutions: taking hold of the reins	41
5	All that glisters . . .	55
6	A cuckoo in the nest?	68
7	'. . . into Parliament he shall go!'	77
8	Arrivals and departures	87
9	'Betsey' no longer as free with her favours	99
10	The ins and outs of office: keeping on track	111
11	The cuckoo exercises his wings	120
12	Royal imprimatur	133
13	Slings and arrows	143
14	Home, sweet Home?	157
15	Meanwhile, back in the colony . . .	174

16	Telegraphs, Honours and the hard slog of government once more	185
17	The Rock	196
18	Career and reputation in tatters?	208
19	A lonely apotheosis: raw young emigrant become Grand Old Man	218
20	Death and resurrection	234
	Appendix A: Weights, measures and money	247
	Appendix B: The Westminster system	250
	Appendix C: The fate of the SAMA	252
	Bibliography	254
	Index	260

ILLUSTRATIONS

Map of the Province of South Australia	ix
Ayers–Breaks–Potts–Lockett Family Tree	x
Rymill–Graham Family Tree	xi

Illustrations between pages 124–125

1. Henry Ayers sketched at 19 at the time of his wedding, in 1840
   (Courtesy of Ayers House, Adelaide)
2. Drawing of a young Anne Ayers (née Potts)
   (Courtesy of Ayers House)
3. Daguerreotype of Henry Ayers, 1847 or '48, taken at Burra Burra
   (SLSA: PRG 67/52)
4. Anne Ayers, photographed during a visit to Melbourne, probably in early 1867
   (SLSA: B 69225)
5. The mature Henry Ayers
   (Courtesy of Ayers House)
6. The mature Anne Ayers
   (Courtesy of Ayers House)
7. Anne Ayers' brother, Frank Potts, purportedly in 1863
   (SLSA: B 45437)
8. Elder daughter Maggie Ayers – 'cause' of the Rock's naming – at the age of 16 in 1864
   (SLSA: B 62969)

9 The Ayers' second son, Harry, at the head of his favourite pony at the age of 12 in 1856
   (SLSA: B 58676)
10 'Heads of the people' – prominent Adelaideans of the 1840s – sketches by S.T. Gill:
   a) James Hurtle Fisher (later Sir)
   b) George Strickland 'Paddy' Kingston (later Sir)
   c) William Paxton
   d) Richard Davies Hanson (later Sir, and Chief Justice)
   e) Captain Charles Harvey Bagot.
   (SLSA: B 350, 349, 334, 330 and 329)
11 The Ayers' North Terrace house, Adelaide, *c.* 1860
   (SLSA: B 8091)
12 The Burra Burra mine site in 1880, three years after serious operations had ceased
   (SLSA: B 12538)
13 Henry Ayers' friend and client, John Benjamin Graham, in 1863
   (SLSA: B 8786)
14 Ayers acolyte-become-critic Henry Rymill, *c.* 1885
   (SLSA: B 45738)
15 Political cartoon from the *Adelaide Lantern* of Sir Henry, at the height of the 1877 political crisis
   (SLSA: z 741.59942 A228d, image 162)
16 William Morgan (later Sir), SAMA Board member, fellow Unitarian and arch political foe
   (SLSA: B 6989)

Photographs 1, 2, 5 and 6 are published courtesy of the management of Ayers House Museum, Adelaide.

Photographs 3, 4 and 7–16 are published courtesy of the State Library of South Australia.

Map of the Province of South Australia

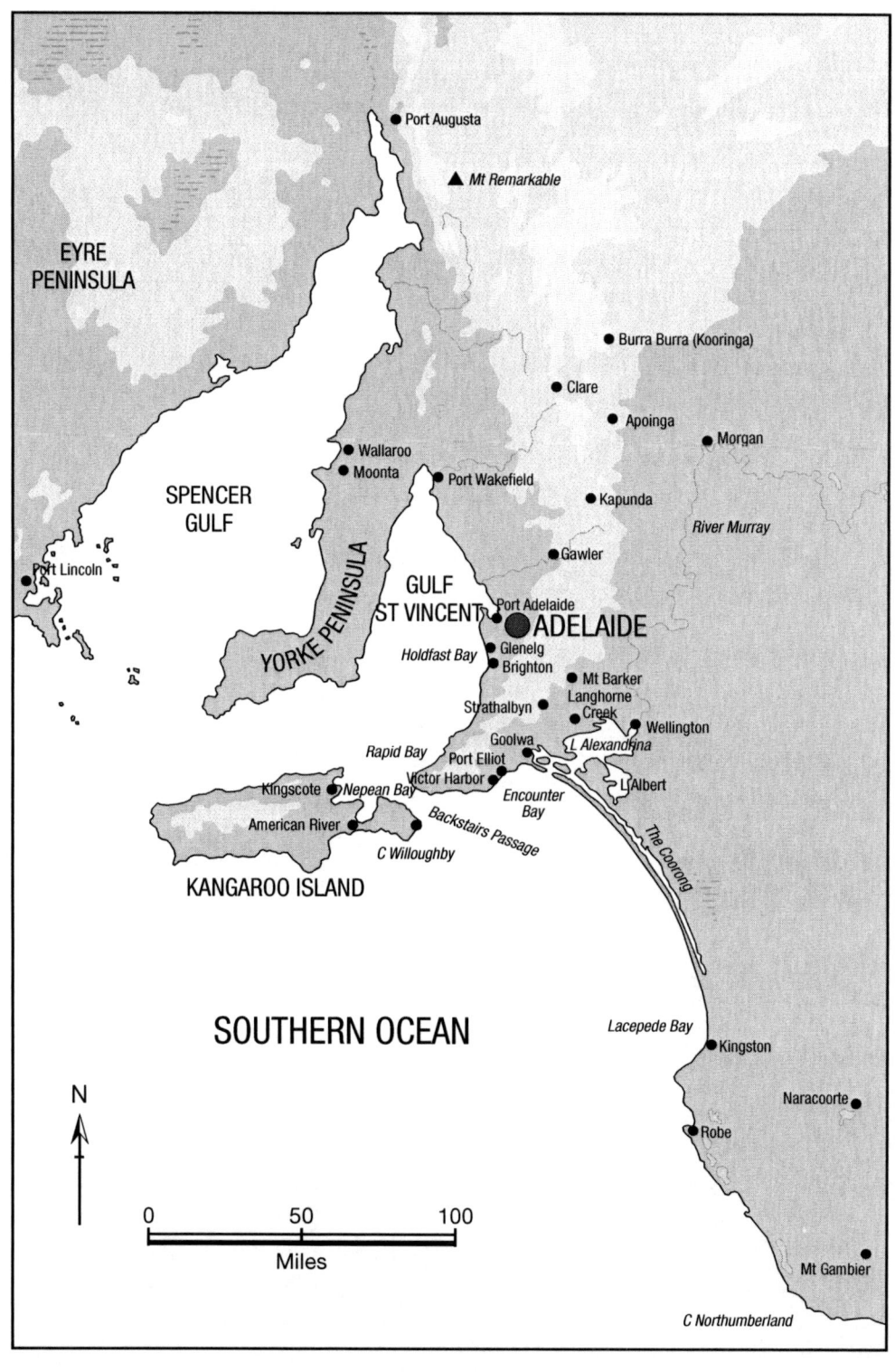

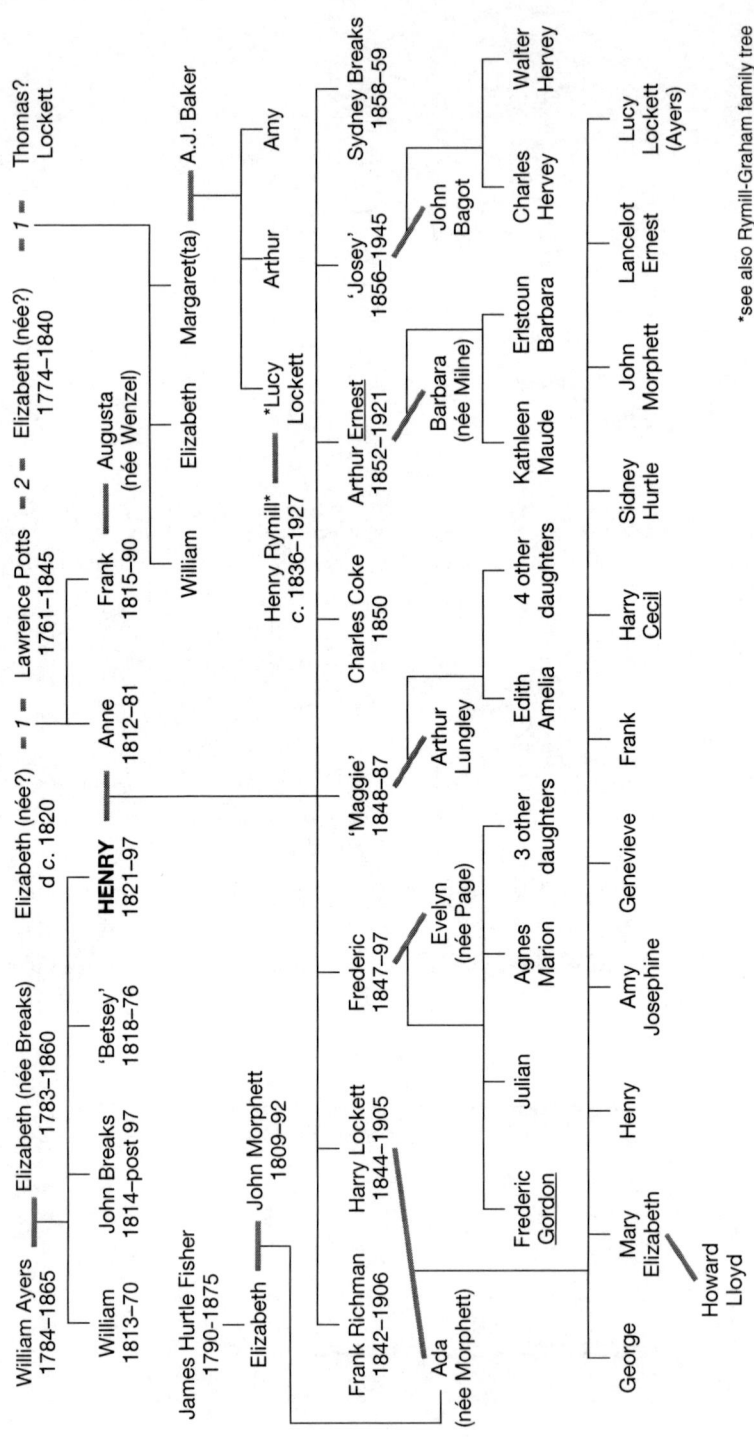

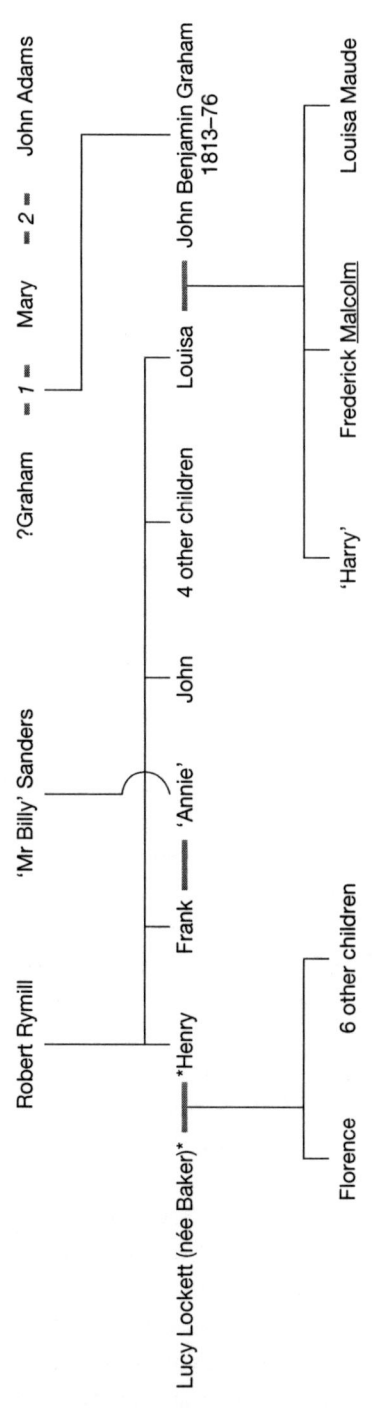

PREFACE

'Who put the Ayers in Ayers Rock?' Once that question struck me, the answer was easy enough to discover. Of more interest, though, was the next question: 'Who was the Ayers behind the Rock?' Like the road to the magnificent monolith itself, my journey towards answering this second question was lengthy.

The initial, unconscious, step involved migration from Britain to Adelaide in the early 1990s. Then came a challenge from fellow members of the Cambrian Welsh Society of South Australia that, of course, I would know of the 'Burra' copper mine, to Adelaide's north. I didn't, but this was swiftly put right. As a family, we explored the surviving dugouts in the banks of the Burra Burra Creek in which, had we been visiting 140 years earlier, we would have encountered Cornish, Welsh and German families who were keeping the wheels of industry turning in the mine; we also wouldn't have failed to notice their smelly herds of domestic livestock. So celebrated was the Burra Burra name in that period that it would spawn like-named imitators as far apart as Devon and Tennessee. How could I not have known about it? I had always had an interest in history and my birthplace lay a stone's throw from the epicentre of the mid-nineteenth-century copper trade. Indeed, around 1860 my great-grandfather had started his working life, as a boy, in the Welsh copper industry.

With the zeal of a convert, I began my own 'mining' project. I scrabbled for any information about this unsuspected British–Australian link that I could win from the depths of the South Australian State Library and State Archives. I scoured the musty ledgers and cash books of the smelting company, based in

'Old' South Wales, which had brought its expertise to the source of such a rich copper deposit as Burra was. I came across the letters of a man hailing from my own home town, who had migrated 150 years before me, and found him writing – over a period spanning about 35 years – to none other than Henry Ayers. They appeared friends. The Ayers name, in the Australian context, looked likely to be connected with that of 'Ayers Rock' fame. It was, and it was also inseparably linked with the Burra copper mining enterprise.

My interest expanded. I had to find out more about this man Ayers, who had, surely, to be larger than life-size himself. Published information about his life proved to be not only sparse but error-prone and sometimes less than flattering. I slowly came to the realization that perhaps the task of trying to discover the real man behind the Rock had fallen to me.

In telling Henry Ayers' story, I have kept to the old money, i.e. to pounds sterling, shillings and pence, imperial weights and measures, and dimensions in miles, feet and inches. Your descent into the mine would have been measured, as with the ocean, in fathoms. If the reader is only comfortable with the metric system, or needs a refresher course on the old system, it might be worth reading Appendix A where I have also set out possible comparisons with today's values for earnings and prices.

Britain is always Home, and alternative homes, such as South Australia, are always home. I have kept contemporary spellings, such as 'honorable' and 'favorite', and even misspellings when quoting directly from letters or newspaper articles. I have retained the underlining of words and phrases in quotes, too.

As Henry becomes active in politics, if you the reader are not familiar with the Westminster system of parliamentary government, it would be useful to read Appendix B before you begin Chapter 7 and, if at the end of his story you have even a tiny yen to know what was the eventual fate of the South Australian Mining Association, the brief Appendix C is for you.

Family trees for Henry Ayers and J.B. Graham are also provided for reference.

Jason Shute
Adelaide

ACKNOWLEDGEMENTS

I would like to express my gratitude to the following people:

My wife Rosalyn for enabling me to chisel away at this work for several years as a 'kept man', and for employing her unsentimental editing skills, as a published author in her own field.

My son Steffan for prodding me to attempt this project, encouraging me as the size of the task sank in, and commenting on the first complete draft. His knowledge of the worlds of law and high finance were also very beneficial.

My daughter Elen, who cheerfully tolerated my recounting of the latest research revelations. She commented on a portion of the text when time allowed during an expedition to the Simpson Desert seeking marsupial moles, which spend their entire time underground!

My son's partner, Bich Anh, and her brother, Minh, who saved my bacon a couple of times over computer problems.

Two Ayers great-grandchildren. The late John H. Bagot, when well in his nineties, was full of interesting reminiscences, particularly of his grandmother Josey – Lucy Josephine Bagot (née Ayers) – or 'Black Granny'. Joanna Catois in Paris, aged 98, was also ready to impart lively memories of her mother, Amy Josephine Cowle (née Ayers), and Amy's elder sister, Mary Elizabeth, Sir Henry's eldest grandchild. Joanna also had touching reminiscences of her grandmother, Ada Ayers (née Morphett).

John Ayers, Ian H. Lloyd and their cousin Gerald S. Hargrave, descendants through Ayers' son Harry and his wife Ada, for their help and encouragement.

Gerry took a particular interest and encouraged me at all points. Through his invitation, I was guest speaker at the commemoration of the 110th anniversary of Sir Henry's death, in 2007, when John H. Bagot memorably laid a wreath on his great-grandfather's grave; since John's death, his son Charles has continued to take an interest in and support the project.

Francis Carpenter, Joanna Catois' son, keen on history in general and keen that the life story of his eminent forebear at last be told, and through whose indefatigable efforts the project has been brought to its conclusion. Francis' cousin James Knapp and his wife Dolly have also been kind in support, in memory of both James' mother Josephine and his aunt Geraldine, elder sisters of Joanna Catois, all three the daughters of Amy Josephine Cowle (née Ayers).

Ross Mathews, for facilitating my researches among late-appearing Henry Ayers papers in the archives of Evans and Ayers. These included Henry and Anne's marriage certificate, and information about the family's early days in Adelaide, clearly not available when Lucy Lockett Ayers compiled her biographical note in 1946.

Gerald Hargrave's sisters, Susanne Newsom and Marianne Davidson, who entrusted me with Ayers papers in the Hargrave family after Lucy Lockett Ayers' days. Also, to Marianne's son James who, with no prompting whatsoever from me, corroborated what had only been my speculation regarding the naming of Uluru 'Ayers Rock'.

Mark Allen, a descendant through Fred Ayers, for maintaining his interest in, and support for, a biography of Sir Henry.

John and Glenys Carpenter, current owners of the Store and storekeeper's cottage at the Burra mine, for their welcome, and to Rod and Colleen Nelson, owners of Thomas Williams' cottage, similarly. Old family friends, the Wrights, of Sauchie-by-Alloa, confirmed that bits of 1846 brick scattered in the Store yard, marked 'Clackmannan', would indeed have come from their area.

Dr Bill Jones, of Cardiff University's History Department, for his encouragement and the supply of information, particularly about H.W. Schneider, and for locating information about Edward Stephens in the Melbourne archives of the ANZ bank.

Staff of the State Library of South Australia who have been most helpful over the years. Those I know by name include Neil Thomas – who has his own family connection with Burra. His knowledge and conversation have always been helpful and encouraging, while his erstwhile colleague, Roger 'the Conservator', kindly offered me study of the newly acquired Coke Papers, prior to his specialist ministrations. To Jan Jones, over an elusive *Lantern* cartoon, to June Edwards over the H.L. and A.E. Ayers Papers, to Andrew Piper and, particularly, Brian Bingley who assisted greatly in the process of my obtaining images from the Collection for the book's illustrations.

Marilyn Jones, Local Studies Librarian at Swansea's central library, who supplied answers to questions regarding the mid-nineteenth-century Swansea–South Australia links.

The always cheery Hannah Phillip – formerly the National Trust of South Australia's Manager of Ayers House Museum, Adelaide, and her volunteer guides – especially for their conversation on things Ayersian over morning tea while I studied the Ayers letters of 1870. Hannah's successors Elspeth Grant and Apryl Morden have also been most helpful. The House has been very kind in allowing use of Collection images which are itemised in the Illustrations list.

Ben Storer and Grace Moore, at Uluṟu Media, for their consultations with the Aṉangu custodians of Uluṟu, in my endeavour to ensure that references to, and the depiction of, 'Ayers Rock' – 'whitefella' history – are not disrespectful to Aṉangu sensitivities. The consideration given by two senior women, Judy Trigger and Barbara Tjikatu, is greatly appreciated.

Helen Bruce, Reserve Archivist of Adelaide University, for her ready help.

Robert Fisher, Archivist at St Peter's College, Adelaide, for information on the Ayers boys' records.

Dr Chris Reynolds, Adjunct Senior Lecturer at Flinders University's Law School, for his knowledge regarding the rights and responsibilities of Aborigines under South Australia's Constitution of 1856.

My good *Freund* Peter Reeh, President of the Adelaider Liedertafel 1858, for locating Handschuhsheim and for confirming its spelling out of a possible three variants I had come across.

Leslie Roberts, expert on that other son of Portsmouth, Charles Dickens.

Graham Williams, who liaised with the South Australian Grand Lodge of Freemasons, facilitating my research among early South Australian Masonic records.

Staff of the Lands Titles Office, Grenfell Street, and of the Old System Section, Carrington Street, Adelaide.

Mel Davies, a fellow Welsh Australian, whose Economic History papers on the Burra enterprise have been very helpful to me.

Staff of both the Old Town Museum, Hastings, and Hastings Museum itself for their helpful background information to the Ayers' visit to Henry's brother John, in 1870.

Richard Parry, a descendant of Henry Ayers' eldest brother, William, for his interest and for information he contributed regarding the wider Ayers family.

My neighbour, Ian Beckingham, for always asking about my latest expedition to the State Library and for matching my daughter's apparent enthusiasm for hearing the latest nugget.

Staff of I.B.Tauris, including Iradj Bagherzade for his initial interest, and for the work and advice of History Editor Liz Friend-Smith, and her successor Joanna Godfrey, and on the administrative side, Maria Marsh, Tatiana Wilde and Liz Stuckey, Victoria Nemeth and Katherine Tulloch. Gratitude is extended further to Sara-May Mallett and her colleagues at Pindar New Zealand, for their input at the production stage.

Finally, I am inestimably grateful to the following, whose generous support has helped enormously towards making this biography possible. (In alphabetical order) descendants of Sir Henry Ayers: Mark Allen, John Ayers, Charles Bagot, Francis Carpenter, Joanna Catois, Gerald S. Hargrave, James Knapp and his wife Dolly, Ian H. Lloyd and family, and Jane Mackinnon (née Ayers). Also to the firm of Evans and Ayers, and their managing partner, Andrew Jonats, similar gratitude is owed. I must further record my gratitude to History SA for its South Australian History Fund grant in support of this publication.

If I have inadvertently omitted to acknowledge anyone, my sincere apologies. In addition, responsibility for any errors is entirely mine.

INTRODUCTION

It might be thought that the Ayers story is a purely Australian one, given the strength of the name's connection with Uluṟu. Certainly, it takes us from the beginnings of a struggling colony in Australia – one without convicts – to the brink of Australian Federation. However, it is also very much a British story, as well as a tale of political and industrial intrigue, office rivalry, and family joys and heartbreak.

Global financial crises also feature. Australia's weathering of the early twenty-first-century version has been largely credited to the country's resource-rich base. The 'soft landing' has been provided by the continued voracious demand for raw materials from a growing and powerful nation – China. In Henry Ayers' day that country would have been Britain. So it was, more than a century and a half ago, for one of the 'proto-Australias', the very young British colony of South Australia, which was foundering at its very outset. 'Gold!' would be the cry just a few years later in the colony of Victoria, next door, but in the 1840s – coinciding with a fevered expansion of railways back Home – the discovery of copper in South Australia was no less tantalizing, not only for the migrants, but for the magnates of Britain's Industrial Revolution. Copper mania was itself a virulent contagion.

'Australia's first mining industry giant' was how Henry Ayers was hailed, as inaugural President of the Australian Institute of Mining and Metallurgy towards the end of a long life in the South Australian homeland he had chosen as a young man. A giant, then, in the life of the 'Monster Mine' which, following discovery of the ore in 1845, was soon 'saving' the colony by alone satisfying a twentieth of the world's copper hunger. Copper for locomotives, copper for sheathing the bottoms of wooden ships and, as that use fell away, the need for copper wiring for telegraphy meant a rising demand for the commodity which South Australia could provide in abundance. With Henry as the steering force behind the Monster Mine, the colony could see itself throwing away its training-wheels and usurping its parent's trade with India and – yes – China. In its raw-material wealth, it rivalled Cuba and a growing Chile. This little Britain beyond the waves – showing scant regard for the Aboriginal owners – was designed to be a non-convict, non-establishment choice for respectable young emigrants. Respectable, though from modest beginnings, Henry Ayers chanced all in the migration adventure which would make his fortune. A powerful figure in the copper industry, and a shrewd financier in his own right, Henry then emerged onto South Australia's political stage, once such a thing had been built as the colony came of age.

A patriotic South Australian by the 1860s, and rising to the position of Premier, Ayers seized the chance for the colony's expansion across the continent to the north. He was determined to deny the neighbouring colonies the chance to consign South Australia to a backwater, should they grab the coming international telegraph line – the internet of its day. This man who loved science, and the advances it brought, would throw whatever resources he could at this transcontinental project, sputtering as it was when he came back into power in 1872. On several occasions, when the young colony faced political crises, it was Ayers who would pull together a government where others could not.

In tracing Ayers' life story, we are given an insight into the experiences of some of those Britons who decided, in the 1830s and '40s, to exchange an uncertain, and probably unhappy, 'Dickensian' existence for a future in South Australia. While still a risky option, it seemed a path worth attempting, offering as it did the promise of life in a new and – as they believed – well-planned society.

Through Ayers' story, we are also brought face to face with re-evaluating the cultural and political twists and turns of our own day. So many issues that seemed settled in Ayers' time have continued to resonate in the early twenty-first century as subjects of open debate.

For example, now as then an innate Australian sense of defencelessness and insecurity means that the powerful friend – in Ayers' day, of course, Britain – must be courted at all costs. At the time, this was reflected in the extravagant loyalism shown in the welcome organized by Ayers for the first ever royal visit to Australia, with scant local acknowledgement of murmurings of republicanism at Home. Private–public partnerships were well understood, Ayers favouring lightly regulated financial institutions – relying on the close scrutiny of shareholders and investors – but as he was by then himself a financier, he would, wouldn't he? Privatization, if not then a recognized term, was certainly an accepted concept. Whether trade unions were a good or a bad thing was debated, as were women's rights. Protection versus free trade, ethics in the business and political spheres, a role or not for foreign investment, how to treat foreign workers, the separation of Church and state, all were big issues of the day, and Ayers would have been able to give you a lively discussion on any of them.

The growth of the idea of Australian federalism, with debates on its desirability, also spanned the majority of Ayers' life as an immigrant – nay, an 'economic migrant'– and Ayers represented South Australia in intercolonial conferences which prepared the way. In 2001, Australians were told by the then Prime Minister to celebrate the wonder of federalism at its centenary, although soon afterwards the very same person maintained that, if the country were starting all over again, federation would not be the answer! Though some politicians at the banquet table in the Adelaide Town Hall, in 1867, were willing to trumpet federalism, and even wanted their royal guest proclaimed hereditary Viceroy – or even King Alfred of Australia – Ayers, as Premier, made no such suggestion.

Henry did his level best to prevent his own business interests from being pummelled by several international financial shocks, and the recent global financial crisis would not have surprised him. He would probably have taken it quite philosophically and, in government, would have reached for the same tools to repair the damaged economy as today's leaders have done: government borrowing for infrastructure projects – although, in what little official memory there is of him, he is depicted as a non-borrower. Those who do not learn from history, we are told, are doomed to make the same mistakes and, as I have read and researched Ayers' story, I conclude that that is not far off the mark.

In 1873 the whitefella 'discovered' Uluru and attached the Ayers name to it, giving the now Sir Henry Ayers a hint of immortality. How far his industrial and political contributions would be remembered, in his colony or around the wider federated Australia, beyond his day remained to be seen. However, the Ayers name would continue to resonate around the world, drawing countless individuals to make a kind of pilgrimage to one of the most remote places on the planet. Could Henry, watching the retreating headlands of Plymouth Sound as he farewelled Olde England and his youthful self, have ever imagined what a mark he might make? Never a mark such as this, surely?

I

'IN SEARCH OF A PLACE TO ASCEND'

Sunday July 20 – Ayers Rock. Barometer 28.07 in., wind east. I rode round the foot of the rock in search of a place to ascend; found a waterhole on the south side, near which I made an attempt to reach the top but found it hopeless. . . . Seeing a spur less abrupt than the rest of the rock, I left the camels here, and after walking and scrambling two miles barefooted, over sharp rocks, succeeded in reaching the summit, and had a view that repaid me for my trouble. . . . This rock is certainly the most wonderful natural feature I have ever seen . . .

Scribbled in his diary, that southern winter morning in 1873, surveyor William Gosse thus recorded his awe as the first white person to reach the central Australian monolith that he named Ayers Rock. Today, the image of its moulded reddish brown form is as familiar to many across the world as the Great Barrier Reef and the more human-scaled Sydney Opera House. Within Australia, the Rock's ancient indigenous name, Uluṟu, has rightly supplanted the nineteenth-century commemoration, but the name of Ayers rings on around the world well over a century after the 'discovery' of the Rock. Yet how many of us have ever paused to wonder what sort of man was he after whom this stunning landmark was named? Was he – could he have been – someone who measured up in any way to the giant accolade accorded him?

The signs are encouraging when one discovers that Henry Ayers served in the highest elected office of the South Australian polity, as its Premier, no fewer than seven times. An easy task, then, one would think, to find a solidly bound biography of the man, to discover his achievements, if only through the tinted lenses of a late nineteenth-century hagiography – but there is none. Ayers' granddaughter, Lucy Lockett Ayers, intended to write his biography, but bequeathed us only a little printed pamphlet of half a dozen or so pages in the 1940s. Otherwise, Ayers is largely – one might almost say conspicuously – missing from accounts of Australian history. As late as the Depression years of the 1930s, his memory could be invoked by the press as an inspirational model for the young generation facing challenging times, but soon afterwards, memories of the man faded. The name of Ayers became simply synonymous with the Rock.

Yet, he was a complex character who played an integral and leading role in the development of the then British Province of South Australia. Born on May Day, 1821, his journey to high political office began with migration as an artisan from his native England in 1840. How did this emigrant, who described himself as a carpenter, come to be a serial Premier of South Australia, President of its Legislative Council, knight of the realm and respected ambassador for his colony's interests at several of the intercolonial conferences that paved the way towards Australian federation? What, in fact, led him to leave his native shores in the first place?

Aiming for 'Improvement'

It would have taken a brave man to commit himself to the adventure of migration to the fledgling colony of South Australia, which was a mere four years old in 1840. Perhaps the impetuosity of youth was Henry Ayers' catalyst, for he was only 19 when, with his new wife and in-laws, he took ship for the long passage. The sea, though not exactly in his blood, was a near neighbour as he had been born and raised in the briny air of Portsmouth, a British naval base. Well within Ayers' father's memory, Admiral Nelson had joined His Majesty's Ship *Victory* from the town itself, for his determined pursuit of the combined French and Spanish fleets.

Ayers' father, William, was a shipwright who married Elizabeth Breaks. His parents' longevity preoccupied Henry in later life, even calculating their average age at death, namely '79 years 78 days', with the parental generation overall reaching an even greater average of '84 years and 264 days'. He seemed to be desperately seeking guidance, in his maturity, about his own life expectancy. Well he might have, because the record of his own generation was far less encouraging.

William's status might have been no grander than that of a skilled artisan but he recognized education as the best route to success in life. He ensured with whatever means he had that Henry, though not his eldest son, enjoyed at least a minimal education. Henry entered the Beneficial Society's school the day before his sixth birthday. Doubtless it was the three Rs that were inculcated. Like the Army and Navy, the school used flogging as an 'encouragement' to its more reluctant pupils, boys receiving their punishment embracing a pillar of the Georgian-style building. The school was not without its forward-looking methods, though, as a pupil-teacher, or monitor, system had been introduced just before Henry's enrolment, known as the 'Madras' System. Carrot, stick or both must have been effective because Henry was sufficiently advanced to be able to take up a position in the Portsea office of S. Blyth, solicitor, at the age of 11 years.

These were not exciting times, though, in Portsmouth. The Navy was in the doldrums and the Napoleonic War a decade in the past. Since the war, high unemployment in Britain had encouraged unrest. The ordinary people were seeking ways of righting some of the wrongs they felt they had to bear due to the oppressive impacts of the Industrial Revolution, then gathering pace. The ideas of that novel breed, economic theorists, were being lapped up by a working people thirsting for knowledge; as Asa Briggs (1959) reports it, the politician Henry Brougham observed: 'The schoolmaster has been abroad in the land.' Religious Nonconformists, and some Evangelical Anglicans, encouraged this aspiration for an improved lot in this life as well as, perhaps, in the next.

Not until Henry Ayers was a toddler did that gloomy era start to give way to one embracing a more relaxed outlook. Just before Henry started school, Britain experienced its first cyclical boom period. Inevitably, bust followed, with unrest simmering again. In 1830, George IV's brother, William, took to the throne and with the new king came a more forward-looking government.

The new reign began at a worrying time, though, with the French having embroiled themselves in another Revolution. Uncertainty across the Channel reignited unrest in Britain, and the new government had to offer the enticements of reform. As Henry stepped into the world as an employee – or employé, as he would have written it – Parliament passed the First Reform Bill, somewhat enlarging the number of the voting public. However, it would be another half century before the majority of adult males could vote, and almost a century before women could do so, at least in Britain. Henry Ayers would see that happen sooner, elsewhere. The 'Great' Reform Act was passed, but it was a tepid compromise. The 1830s, Henry's teenage years, were increasingly a period of unrest and agitation. A run of bad harvests, coupled with commercial recession stemming from the United States, resulted in high living costs and high unemployment. The industrial north, in particular, was badly affected. A million were without work and facing the hated workhouse, systematically introduced to be as unpleasant as possible, to deter the supposedly work-shy. Jobs, when available, entailed virtually unlimited hours and even the regime of a ten-hour day for six days a week was not enacted by Parliament until 1847.

Political reform had fallen short and some idealists worked from newer forums to try to effect 'Improvement' – the goal of the age for all but the most reactionary and complacent. Furthermore, certain thinkers turned their attention to the possibility of creating a new, fairer, though still recognizably British, society overseas.

At 15 years of age, in 1836, Henry found himself in the very place where a ship was being readied to carry a body of early emigrants to the brand new colony of South Australia. HMS *Buffalo*, a naval store-ship, was being converted for this pioneering voyage in Portsmouth dockyard. The Governor-to-be, Captain John Hindmarsh, was also to command the vessel on the lengthy voyage. He had been at sea, man and boy, and was used to expressing his wishes in suitably salty language, which offended the ears of some of the pious, middle-class family members who shared the Captain's Table. The *Buffalo* put out into the Channel in August 1836, with a good proportion of the emigrant party from the Portsmouth area. Whether Henry then knew his future wife, Anne Potts, is unknown, but we do know that her brother, Frank, was aboard as an emigrant.

Frank Potts, at 21, was a retired seafarer. He had trained from the age of nine on *Victory* herself and at 14 had sailed aboard HMS *Challenger*, witnessing the raising of the Union flag inaugurating the colony of Western Australia. He went on to sail Arabian and Indian waters before returning Home, via Australia, the Pacific islands and Cape Horn – passing the *Beagle* carrying Charles Darwin in the opposite direction. With the *Challenger*'s crew paid off, Frank stayed with his parents, at St Mary's Street, Portsea, working as a tallow chandler.

Perhaps this was comfortable enough for a while, but after three years he took the emigration plunge and became a South Australian pioneer, working initially as a crewman for the colony's Harbour Master. Flushed with the enthusiasm of a thrusting young colonist, Frank must have sent the message Home that prospects looked good.

The 'systematic' migration theorists, such as the celebrated Edward Gibbon Wakefield, had wanted to avoid the taint of convictism associated with the existing Australian colonies. They looked to the creation of new homelands for serious-minded, thrifty, hard-working, young married couples, who could go forth and multiply, freed from the backward-looking establishment of the Mother Country. With a modest capital, they could buy land for a reasonable sum, work it steadily and hope to improve their holdings. They could be assisted by labourers who might amass sufficient funds to acquire their own smallholdings and thus move up the social ladder. In theory, at least, the only lands to be purchased were those with which the local Aboriginal population was happy to part. The money from the sale of these 'Crown' lands would fund the emigration of further struggling families. Where the numbers of young couples fell short, the difference should be made up by marriageable youngsters in roughly equal proportions of males and females, the high-minded theorists believing that the preponderantly male 'convict' colonies encouraged all sorts of sinfulness. These young emigrants were to be God-fearing, to meet the expectations of the religious idealists, found in some numbers alongside the purely commercially minded colonizing theorists.

A vast tract of land suitable for this new colony had been earmarked. South-central Australia's coast had been charted by the Royal Navy's Matthew Flinders some 30 years earlier, and the area had since been visited by sealers and whalers, some from as far away as the United States. These had over-wintered in the area, sometimes upsetting the local Kaurna and Ramindjeri peoples by absconding with some of their womenfolk to Kangaroo Island. More recently, Captain Charles Sturt had made his way from New South Wales exploring the length of the mighty River Murray system, hoping to prove it a viable communication link between the east coast and the lands bordering these sealing and whaling waters.

Wakefield and his friends at Westminster, such as the Radical MP, George Grote, put up a parliamentary bill which saw the formation of the Province of South Australia in 1834. It is ironic, for these forward thinkers and frustrated radicals that, as the proclamation of the Province happened at the very end of William IV's reign, the capital was politely named after his queen, Adelaide; she was certainly believed to have strongly opposed recent reforms. The colonizing enthusiasts were anxious to get the venture under way and preparations were supported particularly by Tyneside entrepreneur George Fife Angas, a staunch Baptist who saw the advantages for Nonconformists of living and flourishing in a community free from the 'establishment' influences of the Anglican Church.

Possibly, the Potts family took little persuasion to follow Frank, with the bad harvests in Britain, recession and the country awash with agitation. This disquiet focused itself in 1838 in the movement symbolized by the National Charter, a wish-list of reforms, such as universal (male) suffrage, voting by secret ballot and the abolition of the requirement to be a property owner in order to stand for Parliament. Another point was payment for MPs, which would have enabled a person from society's lower echelons to take a seat. Henry would have a view on politicians' pay.

When the Chartists' demands were rejected by Parliament in 1839, it fell to Sir Charles Napier to quell unrest in the north. Napier had ruled himself out from becoming South Australia's first Governor. Briggs (1959) quotes Napier, who, harbouring sympathy for the oppressed masses, found his duty unpleasant: 'Would that I had gone to [South] Australia and thus been saved this work, produced by Tory injustice and Whig [read proto-Liberal] imbecility;' adding: 'the doctrine of slowly reforming while men are starving is of all things the most silly: famishing men cannot wait'.

Late in life, Henry Ayers gave a celebrated address to the Australian Natives' Association, still delivered in the brogue of his native Hampshire, with his dropped 'h's and country burr. In it, he outlined some of the attitudes prevalent among young prospective colonists in the later 1830s. In his pencilled notes, he observes that many young couples hastened their marriages in anticipation of emigration, and so may it have been for Henry and Anne:

> There is no better period at which a man may commence a great or important or an arduous undertaking than when he has the enthusiastic help and sympathy of a loving wife. It is a time when new enterprises may be undertaken with better hope of success, for man and wife are possessed of double strength with twice the amount of hope and exhilaration than they have at any other era of their existence.

What induced Henry to leave his office position and risk life, limb and career in ocean voyaging and colonial pioneering is nowhere made clear. Coming of a family that was at least nominally Anglican, he was apparently not fleeing the oppression of an uncongenial established Church. Perhaps there was a deadmen's-shoes situation over promotion at work. He certainly later acknowledged the unsettled and unpromising state of Britain as strongly motivating educated young people and those with trades to emigrate.

By 1840, Henry Ayers had determined on a life in South Australia. A positive impetus may have come from his prospective in-laws. Lawrence Potts had brought his young family to Portsmouth from Hounslow some time between 1816 and 1820. As a widower with an eight year old, Anne, and Frank at five, Lawrence had married Elizabeth Lockett. Elizabeth appears the breadwinner, as a milliner and dressmaker, with Lawrence, somewhat getting on in years, no longer practising linen-drapery. They were living in much the same part of town as the Ayers family, but how Henry and Anne met and resolved on a life together on the other side of the globe must remain a matter of conjecture. Elizabeth Potts' elder daughter, also an Elizabeth, had married John Churcher in 1830 and, by 1838, her younger sister Margaretta (Margaret) had followed her down the aisle. Stepsister Anne was to follow their example in not marrying young.

An 1869 copy of Henry and Anne's marriage certificate shows that elderly father-in-law Potts was by 1840 carrying on the light trade of bookseller. The certificate answers a few basic questions, but also deepens a mystery: was Henry and Anne's marriage, at the outset, a love-match at all? The age disparity is somewhat unusual; he was just 19, she getting on for a decade older. The certificate dates her baptism at New Year, 1813, and indeed the Hounslow register records her birth as 28 November 1812. Many years later Henry noted her birth date as November 1813. Had Anne knocked a year off her age in an attempt to minimize the difference between them, or was she herself simply mistaken? Did Henry go along with Anne's accepted age, reducing the gap for the sake of his vanity? Perhaps he thought a slightly smaller gap would help to overcome any family disapproval; he was the 'baby' after all. In his application to the Emigration Commissioners for a free passage, he gave his own age correctly, but lopped a few more years off Anne's, claiming for her a rather generic 21. On the copy of the marriage certificate, they are both noted as simply 'of full age', which of course Henry was not.

The other question thrown up by the certificate concerns the wedding venue. There were several churches on the east side of Portsmouth harbour, where Henry worked and both families lived, including St Mary's, Kingston with which both families were associated. So why did they marry in St Mary's, Alverstoke? This church was well to the west side of the harbour, beyond the imposing Fort Blockhouse. A possible, though very faded, signpost to this venue is the fact that Alverstoke was home to Churchers, of whom there were very few in the rest of Hampshire. Anne's stepsister, Elizabeth Churcher, had not yet emigrated to Canada, following her sister, Margaret Baker and brother, William Lockett; she witnessed Henry and Anne's marriage. Possibly, a Churcher family connection enabled the ceremony to take place in Alverstoke. At all events, the circumstances surrounding this wedding do not seem quite normal.

Was it initially a marriage of convenience? If the Potts' main aim was to reunite with Frank – Elizabeth Potts wrote revealing such a desire – it made sense for Anne at least to get a free passage as part of a needed, young, productive married couple. Perhaps she even instigated, and drew young Henry into, the plan. From Henry's viewpoint, as he would not qualify for a free passage as a lawyer's clerk, he had to convince emigration agents that he was skilled with his hands. What better way to prove this than to have it in black and white on the marriage certificate? Henry Ayers was a 'carpenter'. With father William a shipwright, and eldest brother William himself a carpenter, Henry could probably talk at least a little 'carpenterese'. St Mary's, Alverstoke, may have been sufficiently isolated from the Portsea world of the Ayers and Potts families as to allow the best chance of success for Henry's subterfuge.

The incumbent of St Mary's, Kingston, might have done more than just raise a quizzical look had Henry tried his gambit there.

The State of the Colony

The Potts and Ayers couples received news of the young colony from Frank but, since it generally took four or five months for correspondence to travel back Home, they could not know that, by the time they set sail in July 1840, South Australia's prospects had turned quite ugly. The theories of the Wakefields, Angases and Grotes had failed to work as efficiently in practice as these Systematic Colonizers had hoped. Difficulties in preparing city and country land for sale, while also seeking other sites for settlements and harbours, meant the first colonists were slow in establishing farming. Furthermore, the theorists had not accounted for problems arising from land-purchasers who had no intentions of experiencing personally the rigours of a pioneering life, preferring to have agents do the task or, worse still, do nothing and wait for the property values to rise before walking away with the profits. In many cases, the earliest colonists were forced to remain in the security of the nascent city of Adelaide, where land speculation quickly absorbed their interest and, often, their funds. Fortunes, even modest ones, could be lost as well as made. At several moments, whether the new colony could actually feed itself was in doubt. Governor Hindmarsh faced dissatisfied colonists and rubbed up regularly against his co-authority figures, Resident Commissioner James Hurtle Fisher, and Surveyor-General Colonel Light.

Hindmarsh's performance caused dissatisfaction, both within the colony and at Home, resulting in his recall. His replacement, Lieutenant-Colonel George Gawler, resolved to keep the South Australian venture afloat at all costs. Fairly large-scale public works were undertaken, not only to turn the Government Hut into a Government House, but also to otherwise adorn this rough and ready young capital. When Gawler's lavish spending rang alarm bells at the heart of the British government, which would be footing the bill, it became clear that he, too, would have to be recalled; it was by then too late for second thoughts by Henry, Anne and her parents.

The shock of insolvency hit the colony. With the Adelaide Treasury empty, the first signs of unemployment were recorded in June 1840, the same month that Henry abandoned the Portsea solicitor's office and took the ferry to marry Anne. After just a month of wedlock, the couple set sail on 15 July, aboard the *Fairfield*, bound for South Australia.

Whether those aboard ever expected to see their home shores again, we cannot know; it would be something for which Henry longingly wished. His, after all, had been the sacrifice of leaving Home, and family, behind. What did his brothers, William and John, and his sister – another Elizabeth – make of the adventure? Were they quietly envious of their little brother's courage and the opportunities that might come his way? One can only guess at the emotions that William and Elizabeth Ayers must have experienced at the departure of their youngest, suddenly a married man, and just as suddenly sailing for an experiment of a colony on the far side of the world.

A Fresh Start

The Ayers and the Potts shared the experience of thousands who voyaged to Australia under sail. The care of passengers was quite closely scrutinized. Medical attention was mandatory aboard these ships and some captains' remuneration was held over as a bonus, paid out once their charges had provided a glowing reference for their paternal qualities displayed during the vicissitudes of the voyage.

By 17 December, the *Fairfield* came to anchor in South Australian waters in Holdfast Bay, Colonel Light's prized anchorage. Frank Potts was undoubtedly able to put out to the *Fairfield* before she even came finally to anchor – perhaps in the boat he had built, *Musquito* – to be reunited with his family after over four years. Even if he had known his new brother-in-law previously, he would only have remembered Henry as a lad. He now met a wavy-haired young man of average height, still with the slimness and wispy beard of youth. But what a mix of emotions Frank must have experienced to find him accompanied by only his father and sister. Despite the comparative safety of the Australian voyage, tragedy could and did strike during the months at sea. On 8 October, short of Cape Town, Mrs Potts had died, racked by seasickness. She was 66. Consigning her body to the deep, it only remained for Anne and Henry to try to comfort old Lawrence Potts in the loss of a second wife.

Henry wasted no time pretending to live up to claims of carpentry skills. Following Christmas celebrations, he began work in the law office of J.H. Richman, on Adelaide's desirable North Terrace. For a while, the Ayers would have to settle for accommodation in less significant parts of town. Lucy Lockett Ayers says their initial address is unknown but, by 1843, they were right in the centre of the 'square mile', in Wakefield Street. Henry could count himself extremely lucky to have secured a position in his true field so soon. There was a surfeit of young middle-class professional men, in a colonial population of about 16,000, which was thinly spread over an area some 100 miles by 50, though still heavily centralized in the city.

Four years of development, even with Gawler's boost to public works, had not raised the capital to any great heights of amenity. New arrivals often suffered from dysentery, as a result of a less than satisfactory water supply. This was sourced largely from the sluggish waterholes in the bed of the River Torrens, which meandered through the shallow valley it had long ago cut, now serving to decorously separate Colonel Light's North Adelaide from his city proper. The Torrens, or as the Kaurna Aboriginal inhabitants knew it, the Karrawirraparri, could sometimes run over-vigorously in winter, but the Ayers had arrived in the first of the hot summer months, which was a testing time for Europeans each year. New arrivals suffered eye afflictions and, as the visitor Edward Lloyd described in 1844, the dry north wind could induce a feeling of fever and the very copper coins in one's pocket felt 'odiously hot'. Lloyd described the red dust from the mere three streets that were reasonably developed, caking buildings, as well as citizens striding purposefully about the town.

The residents of the young metropolis had just been given a chance to make improvements for themselves. Governor Gawler, who personally embodied the powers of British Government representative and Colonial Commissioner, had encouraged the formation of an elected City Council. It is as well to remember that this privilege was then a novelty for many sizeable British towns; some old boroughs had charters dating from the Middle Ages, but many of the newer towns, growing through the effects of the Industrial Revolution, had only just been given the opportunity to form representative bodies.

Perhaps Gawler's timing was not the best, though. The Council was formed just when many colonists were heading out of the security of Adelaide into the surrounding districts, applying what little farming expertise they had in a novel environment of both soil and climate. The city's population was actually declining, then, at a time when a rate needed raising for the Council to be able to make improvements and generally fulfil its obligations. The first mayor was the former Resident Commissioner alongside Hindmarsh, James Hurtle Fisher, whose powers had now been combined with those of Governor Gawler.

Henry had only been in the colony for six months when Gawler was replaced by a much younger army officer, Captain George Grey, who had made a name for himself as an explorer in Western Australia. At 27, his youth was held against him both by the local press and, naturally, by the recalled Governor's family. It was Grey's duty to rescue South Australia from Gawler's profligate regime – as it was perceived by the Colonial Office in Whitehall. Retrenchment was Grey's order of the day, bringing an end to governmental support for any surplus labour.

Hard times ensued for the colonists and Grey's resolve was tested when the hungry unemployed and the bankrupt marched on Government House. With coinage short and land values plunging, Grey plumbed the depths of local unpopularity, although he did do something to alleviate distress by charitable giving out of his personal resources. As 1841 wore on, Henry must have counted himself extremely lucky to have continued in employment at all and, during 1842, numbers of pioneering colonists simply gave up and left. At some point during their first year in the colony, Anne, at least, found prospects dispiriting, and put her feelings into verse addressed to their pet cockatoo:

> I wander'd forth in this new Land to think upon my [H]ome,
> To heave a sigh to absent friends, and ask why did I roam?
> Why leave the peaceful happy vale, and cot that once was mine,
> But leagues divide me from that spot, 'tis useless to repine.
>
> I'll think on this poor captive bird, and learn from him to bear
> The cruel loss of [H]ome, and friends, a fate he seems to share.
> Sweet bird to ease my exiled heart, Ah! say thou dost not pine
> For thy early scenes, and mountain range, that ne'er again is thine.
>
> Happily thou dost forget the past, not dread the coming morrow,
> Nor wear thy heart, in vain regret in hopeless, ceaseless, sorrow.
> Such tiny limbs to bear a chain, humanity 'twould shock ye,
> The gentle creature raised his head, and answered, 'Pretty Cockey.'

Cockey seems a good listener and capable, with a couple of well-chosen words at just the right moment, of taking the sting from Anne's melancholy.

The one bright light was an overwhelming harvest which, even with the assistance of many unaccustomed reapers, could not all be gathered in. Perhaps Henry turned out to help. This is not as far-fetched as it might seem because, over the following two years, Henry owned a small farm, to the city's south-west. Whether he worked on it himself, providing for the family table 'allotment'-style, or had someone work it for him is not clear. Financially, it was not a great success, as he lost almost £60 on the deal when disposing of the property. He had the upkeep of a 'poney', as well, which he also parted with – at a loss – and one can readily picture the little animal's short legs going nineteen-to-the-dozen in carrying Henry down to dig over his plot.

In March 1842, Anne gave birth to their first child; she was in her thirtieth year and Henry a couple of months off his age of majority. Perhaps feelings of apprehension during her pregnancy had contributed to the nostalgic mood captured in her poem to her confessor Cockey. She had been facing the ordeal of childbirth and motherhood without a mother-figure, or even her older stepsisters. She faced the dread 'coming morrow' with only a very youthful husband, and an elderly father, for reassurance. The Ayers heir, safely delivered, was named after, perhaps, the two most influential people in Henry's new life so far – Frank, his brother-in-law, and Richman, in whose service the Ayers had enjoyed some security during this very difficult period.

The following month, Ayers moved into the joint practice which J.H. Richman formed with his new partner, Macdonald, in Wakefield Street. This was a come-down after North Terrace and a good step from the main business streets. This may have had immediate implications for their legal business as Henry notes a more prestigious address alongside the entry in his *Memorabilia*: 'afterwards King William Street'.

Within a year, though, Henry moved to a position in the Stephens Place law practice of James Hurtle Fisher, ex-Resident Commissioner, and now ex-Mayor of Adelaide to boot; Fisher had resigned when an even higher rate could not be raised to pay off the debts that the Council had already incurred. At the following municipal election, a mere 135 citizens had bothered to enrol to vote, making polling pointless. Governor Grey made a feint of trying to rescue the municipality's honour by co-opting influential citizens onto the Council, but he happily acquiesced in London's decision to end the local government experiment, when that instruction arrived by mid-1843. The seeming civic apathy and failure of the Council was not a good advertisement for the colony, boding ill for its chances of successfully achieving representative government, which had already been promised, once its population reached 50,000.

Henry and Anne were blessed with a second son, Harry Lockett, in April 1844. His second name kept alive the former married name of Anne's stepmother, after her sad death at sea. Henry immediately 'hopped' law practices again, becoming employed by Smart, Johnson, and Bayne, solicitors, also of Stephens Place. This would prove the crucial, positioning move in shaping his career, and in deciding the sort of life his family could enjoy.

The Ayers' three and a half years in South Australia had coincided with the poorest conditions and prospects that the colony was to know until the threat to its very survival presented by the gold-rush in neighbouring Victoria. Despite this, industrious Henry had obviously made the best of his opportunities. He augmented his income through a sideline in supplying law stationery. He was also picking up plenty of extra pay through 'overworking' and, by 1844, had built a house in Pulteney Street and was no longer paying rent. Rather, he was *collecting* rent on three properties, one a present from a very generous Frank Potts. Having paid back modest loans to his brother, John, and brother-in-law Potts, the astute Ayers would soon be in a position to lend money, to erstwhile employer J.H. Richman for one. Henry was using the downturn in the economy to his advantage and, all told, circumstances were starting to vindicate the decision to emigrate. Yet even greater opportunities soon beckoned.

2

THE SECRETARY

The colony's fortunes turned in 1844. Governor Grey's stringencies were starting to pay off, and the working of a sizeable body of copper ore about 50 miles to Adelaide's north-east, at Kapunda, indicated a future for the colony beyond wool and wheat. This mineral discovery by a son of local pastoralist Captain Bagot, while out tending his father's flock, must have had a consonantly Old Testament ring about it to the more pious colonists.

Mineral fever ensued. Every shepherd and country bumpkin seemed able to turf out metalliferous pebbles from the deep recesses of a smock or greatcoat, to tantalize the greedy or gullible in town with sufficient capital to underwrite a small mining venture. Nor were speculators at Home behindhand, once the outlook for a return on investment in the colony looked brighter.

In 1845, extraordinary copper lodes were discovered near the Burra Burra Creek by distinctly poor shepherds on 'Crown' land – actually country of the Ngadjuri people. The name of this large watercourse was not in fact Aboriginal, but derived from the Hindi for 'Big Big'. The area was almost twice as far from Adelaide and its port as Kapunda, a critical consideration in realizing the potential of the find. The raw material needed processing with expertise born of long experience, best found within the arc of the South Wales coalfield, where copper had been smelted since Queen Elizabeth I's time. Capitalizing on the Burra Burra finds would call for deep pockets, not only to purchase the land and work the lodes, but also to organize ore transport by land and sea.

Caught up in the mineral fever, a group of reasonably well off city shopkeepers pooled their capital to invest in any newly discovered mineral lands, taking the name South Australian Mining Association – the SAMA. The two Burra Burra finds could be acquired in a single purchase, a 'Special Survey'. These large acreages, well beyond the pockets of genuine small settlers, had been introduced by Governor Gawler to attract major investment from Home into the colony. The Burra Burra land was available only as an enormous 20,000-acre Special Survey at £1 per acre, in cash. It seemed doubtful that anyone in the colony could lay hands on such a sum. Ian Auhl, in his book *The Story of the 'Monster Mine'* (1986), quotes the rhyme doing the rounds summing up the panic:

> The snobs they are hurrying up and down
> Some lend them a sovereign and some a crown
> They have not a shilling in all the town
> To pay for a pint of wine.

He tells of elderly ladies producing what cash they had, secreted in their stocking-tops, to help make up the total.

Pivotally, Henry Ayers now became involved in SAMA activities. To understand his rise in colonial society, in its commerce, its politics and its cultural life, one must grasp the significance of the Burra enterprise and its prominent contribution to the success of an eight-year-old South Australia, still recovering from its first depression.

Ayers himself devised a history of the SAMA, 40 years on, drafted thriftily on scraps of paper, such as the backs of Adelaide Choral Society flyers. His intention was to give some idea to the then citizens, living in something of a 'mining depression', of how the earlier generation had countered bad times and a want of money. It was hoped that the grouping of interested parties would prevent new mineral resources falling into the hands of single proprietors. The SAMA's Provisional Committee first met on 5 April 1845, with Henry, he tells us, as 'provisional Secretary'.

The Association pushed its shares and petitioned the Governor to speedily put up for sale any new mineral lands. A Board of Directors was elected and a 'secret and confidential Committee' deputed to work with the Trustees to purchase certain Crown lands. Several of these gentlemen had recently raised their profiles in agitation to quash the 'Parkhurst Boys' scheme. The colony was threatened with a whiff of convictism, given the British Government's plans to send young offenders to South Australia as supported emigrants with free pardons. The SAMA's new committee-men had pointed out that, if the scourings of Parkhurst and Pentonville were sent, it would be against the letter of the Act which had set up South Australia.

On 21 April, Henry was elected Secretary. Smart and Bayne were confirmed as the Association's solicitors and Henry's close connection with them probably weighed with voters beyond any knowledge of his particular suitability. His duties were initially restricted to minute keeping and document copying, but on his twenty-fourth birthday the first letter bearing his signature as Secretary was issued from Smart and Bayne's premises. Compared with the madcap month of April, during which his family could have seen very little of Henry, May saw a welcome slackening in the break-neck pace of meetings.

According to Henry, it was to him that the shepherd, Streair, brought specimens for inspection from a lode discovered near the Burra Burra Creek, in June 1845. If the bank managers could be satisfied as to the quality of the copper find, funds for a Special Survey might be forthcoming. Henry noted that the Bank of South Australia had just over £20,000 at its command. However, in the light of previous colonial profligacy, as viewed from Home, manager Stephens was on a tight rein from London in terms of speculative lending. Perhaps he authored the letter in *The South Australian Register* urging people not to be distracted from the 'main' aim of colonizing – presumably, agrarian pursuits. Speculation, he added, would only make something of a bad name for the colony even worse. Stephens' competitor, the more recently arrived Bank of Australasia, could only boast a quarter of the first bank's cash on hand.

The dearth of funds forced a temporary marriage of necessity between the SAMA and some wealthy landowners, to fund the land purchase. The landowning 'Nobs' were agreeable to then draw lots with the shopkeeping 'Snobs' for ownership of either Streair's southern lode or the northern lode, spotted by shepherd Thomas Pickett.

The SAMA had promised ample rewards for information about finds, and its directors later became subject to finger pointing regarding a failure to recompense the shepherds whose finds would lead others to great wealth, and the same mud stuck to Henry. Public outrage especially followed Pickett's death through slumping – drunk – face first into his camp fire. Henry indeed tells us that he gave Streair no assurance of recompense, citing the great distance involved, estimated at over 90 miles and a three-day journey from Adelaide. He referred him to Director Bunce, who does seem to have given him something for his trouble. Perhaps Henry's retrospective explanation of his own role constituted self-defence, but it does demonstrate that rewards were a matter for the directors, not the Secretary, which was at that stage a fairly lowly administrative position.

The *Register* described the whole population as 'agog' with the antics, as though 'mesmerised . . . from greybeard to spinster, from sedate matron or spinster of a certain age to sweet seventeen', and carried articles on mesmerism and clairvoyance in the same edition. Not that the editor was disinclined himself to capitalize on the mesmerized; in his premises, where he also sold books, stationery and patent medicines, he added a line of geologists' hammers. His stock of books was expanded, too, with titles to interest mineralogists and explorers.

Following a cursory land survey by William Jacob – later to lend his name to one of South Australia's most popular wine exports – Surveyor-General Frome prepared to do the Special Survey. The Association deputed one of its own members to attend. George Strickland Kingston, formerly less-than-competent deputy to founding Surveyor-General Light, had returned to practising his real calling, as the City Council Engineer, while also dabbling in architecture. After considerable toing and froing between Frome and the pernickety Kingston, the survey was agreed and on 20 September the meeting for the divorce of Nobs and Snobs was held, as Henry later described, at Platts' Rooms, fronting on Hindley Street. If they felt peckish, they could be well supplied with sandwiches and pastries from the little café adjoining, though each probably had a stomach full of butterflies in any case. The lots were drawn. The lucky Snobs drew the desirable northern lode, which became known as the Burra Burra Mine. The Nobs took the booby prize, the short-lived Princess Royal Mine. Henry had good cause for being thankful that his horse had been hitched to the SAMA's wagon.

Right in the middle of this feverish commercial activity, Anne's father died. The *Register* went a little overboard in describing Lawrence Potts as approaching his ninetieth year; he was 84. It was correct, though, in believing him one of the oldest colonists, and accorded him the title 'father of the colony'. The little notice also commended Anne Ayers and Frank Potts for their devoted care during their father's 'decline', of which papa Potts could never speak highly enough. As a dutiful son-in-law, Henry organized and funded the funeral. Most of South Australia's early colonists were comparatively young people, separated from the older generations of their families; so it now was for Henry and Anne.

Exploiting the Prize

The SAMA's Board turned to the task of exploiting its prize. Cornish copper miner Thomas Roberts was engaged to undertake the initial blasting. Before heading north, he received full instructions in the hand of Henry Ayers, specifying in businesslike form the terms of Roberts' engagement. This was the first of many such outward letters to be copied into an enormous new tome, far larger than a family Bible. In Henry's neat, unostentatious hand, Roberts' pay was confirmed at £3 a week. A cart, pulled by two horses, was to convey him, his men and their immediate requirements to the Burra Creek and they would have to ride in turns as the cart would not carry them all. They should call en route at F.H. Dutton's sheep station for tools. On reaching their destination, and if they were not 'too jaded', they should send the cart back for the stragglers. Any delay should be avoided in providing ore for trading; the SAMA knew it would be months before they could expect any return on their outlay, through sheer distance. Tarpaulins would prevent wasted time constructing anything as substantial as a hut. This focus on immediate attention to duty and apparent lack of concern for the well-being of the workforce was to be an ongoing hallmark of the SAMA directors, relayed to the mine officers through their mouthpiece, Henry Ayers. Though he was only the messenger at that stage, the beginnings of a reputation as a hard man were being established.

The transport question was also addressed, through two bullock-drawn drays belonging to J.B. Hack. Henry instructed that tools and stores be taken up and the first few tons of ore brought back – and one of Roberts' carthorses. Hack should also scout for stopping stations for regular runs. While at the mine he should not be idle as the drivers should take every opportunity to help draw wood to assist the miners in constructing, despite previous instructions, their 'temporary' huts.

With mining and carting in hand, a third element needed addressing: smelting. The SAMA engaged German smelters Captain Augustus Ey and Georg Ludwig Dreyer and his son, who were on the spot. Confirming the novice Mining Association's determination to be as independent of overseas smelting expertise as possible, Henry commissioned designs from architect-surveyor-engineer Kingston, not only for cottages for the miners and their families, but also for a fully fledged 'Smelting House'. This was not a great triumph. Kingston failed to adequately consult the engaged experts and his partially constructed smelter, its walls barely risen, had its two furnaces badly positioned, preventing free access to their interiors.

The name of the mining hamlet of original cottages was to have been Redruth, no doubt aimed at giving the Cornish miners some sense of Home, but it was soon renamed Cooringa, or Kooringa, an Aboriginal name meaning 'place of she-oaks'. Kingston's miners' cottages constituted a small-scale example, in a pioneer colonial context, of the forward-thinking attitude of some contemporary British industrialists who thought it responsible to provide model housing for their workforce. However, hundreds would turn down 21-year leases of these attractive cottages, voting with their feet – and a pick and shovel – to dig their own dwellings into the banks of the Burra Burra Creek.

The workers knew only too well that the mine could prove a dud any day, as indeed would be the fate of the Princess Royal Mine. The creek dwellings soon housed a population of some one and a half thousand. Inevitably, the creek became polluted from mining run-off and the general detritus of human activity, including slurry from pig-keeping. Squalid 'Creek Street' was far from the ideal place for families to live, and the SAMA could not understand why so many continued to shun the available housing. The directors also deplored the lack of rents.

The attack on the exposed lode began on 29 September 1845, though the availability of experienced miners was somewhat problematic, with 'poaching' between mines occurring. Thomas Roberts was replaced as Mine Captain by Ferdinand von Sommer, again with a Henry Ayers 'special' by way of a letter of instruction. Von Sommer's letters to Ayers reveal a good grasp of English, tinged with occasional Germanisms, such as a delicious coinage when, asked to send a poor specimen of a grey horse back to Adelaide, he says he will send it 'townwards'. He implemented the 'tutwork' system, or sinking of shafts by miners paid at so much per fathom – vital work, but not immediately productive like 'tributework', the raising of ore on commission. Von Sommer's spelling 'Totwork' is a nice German–English hybrid, signifying 'deadwork'. From von Sommer, we learn further that Kingston had constructed the smelting house furnaces out of non-heat-resistant soft clay, and with no protection for the bellows. Worse still, no access was allowed for the bellows to connect with the furnaces.

In the middle of 1846 Henry's salary was raised to £200 per annum, having risen steadily from his temporary starting salary of £50. The SAMA directors now required him to attend exclusively to SAMA matters. Presumably, he had previously continued some work for Smart and Bayne, where the SAMA leased a room for £10 a year. More formalized office arrangements came with tighter SAMA board procedures, making a Secretary's task easier: discussion was to be of the business only, and there was to be no smoking either.

Henry and his family must again have counted themselves fortunate. Having worked for several Adelaide employers, it was pure luck that Henry found himself with a sufficient level of expertise and experience in the firm of Smart and Bayne when it was chosen as the SAMA's solicitors, providing him with his opening into the thriving new concern.

Teething Troubles

February 1846 brought both success and near failure. Ore was being raised, but the miners' mates were threatening to leave on account of low pay. That valuable auxiliary, the cook, was also just persuaded from leaving; whether his intended departure was from dissatisfaction with his rewards or a lack of appreciation of his culinary skills is not clear. At any rate, von Sommer managed to convince his workers to continue, if only until the visit to the mine of the directors.

Tours of inspection to Burra, as the mine and surrounding settlements became commonly known, were a regular feature and Henry was frequently required to be in attendance. His first official trip came in the October of 1845. Numbers of directors, and Kingston, were going up to see the results of the mine's opening. Henry greatly enjoyed these opportunities to get out of the office and they seem to have been undertaken in a merry spirit by the carriage-ful of directors. When they travelled, the owner of the livery stable, Jamie Chambers, took the reins himself. Comfort was generally absent in these early days, and the journey could be quite hazardous. On one occasion, the party had to cross the swollen River Para in an ox-skin punt. Another carriageful of people, attempting a crossing of the Light a few years later, was swept away; all five aboard, including a doctor bound for Burra, lost their lives.

Apart from the hazards of travel over unmade roads and occasionally swollen creeks, accommodation could be problematic. Initially, the visitors relied on the enforced hospitality of their mine officers, which encouraged an undesirable over-familiarity between master and servant. When writing Home to the mine's largest shareholder, J.B. Graham, Henry expressed his relief at the improvement once the Burra Hotel was opened, late in 1847; he appreciated both the comfort and the independence from the mine's staff.

Before the end of January 1846, the board decided that they needed more direct control of mine operations, with von Sommer too costly to retain. They installed Samuel Stocks Jnr. as their Resident Director. His salary of £400 plus house, water and candles was still not to be sneezed at. The directors' largesse did not extend to the workers, though; the men had been given nothing for their first Christmas Day holiday, nor for their afternoon off on Christmas Eve. Some years later, the secretary of another mining company asked Henry whether the SAMA paid for a Whit Monday holiday. Not much was accomplished on that day, Henry admitted, but added that the Association had only ever paid men for work done, and never paid holidays.

Samuel Stocks, too, carried a letter of instruction in Henry's hand on taking up his post in February. A general merchant and prominent committee-man of the young Mining Association, Stocks nevertheless had little knowledge of mining, and had to rely on the experienced mine captains. He got no more than had his predecessor out of the Dreyers' smelting experiments, and they had their own problems with their bellows mechanism – provided by an Adelaide pipe-organ builder. On occasion, father and son preferred the comforts of town to standing before a glowing furnace, thus breaching their contract. It became clear that a reliance on Swansea, Liverpool or London smelters could hardly be avoided. Henry's duties therefore broadened to include correspondence negotiating terms for shipping and insuring the richer ores.

The board's unhappiness with Stocks' performance quickly showed. Replacement Captain Henry Chipman came more cheaply, at £225, and Henry's own salary nosed ahead in mid-1847, at a straight £250. Little extras included the odd five-guinea fee for attending an audit. Chipman, too, struggled and Ayers eventually had to write a letter of dismissal, referring to his drunken unpleasantness during one of the directors' inspections.

Despite the personnel problems, progress was such that the second AGM, in April 1847, was a 'harmonious', if not 'jocular' affair, according to the *Register*: 'the volumes of smoke from the cigars above stairs being responded to by the jingling of brandy and water glasses below and many a smart joke was exchanged between the teetotallers and those who are only temperance men'. It is safe to include Henry Ayers among the latter, although it is extremely doubtful that he was among the smoking fraternity. Incidentally, within a month of that occasion, the paper published a long list of unpleasant symptoms a Dr Laycock had discerned in heavy smokers. Henry was an inveterate collector of newspaper articles on such health matters, yet he does not appear anything of a hypochondriac; in fact, on occasion, he could be quite careless with his health.

When news arrived in the colony – taking only a record 57 days from London, via Ceylon (Sri Lanka) and Mauritius – that the £5 Burra shares had risen as far as £80 each, it came with a sour-grapes comment that that might still be a little too low for the 'spoiled children of fortune of South Australia'. Better news still for SAMA shareholders was published in the *Register* in July; £5 Burra shares stood at a value of £100, although some might have wished they had shares in the Princess Royal, which had risen to £259!

Burr at the Burra

The directors finally felt they had found the right man to run their on-site establishment, just before the second anniversary of the mine's opening. Thomas Burr was well respected as the colony's Deputy Surveyor-General. The fact that his father was Professor of Mineralogy at Sandhurst College had to be a plus, too. Burr's letter of engagement as mine Superintendent laid down, as usual, his wide-ranging duties, for which he was to receive a very handsome sum of £700 per annum, along with a house, fuel and water. Salaries were on the rise all round and Henry's was raised to £350 – a sizeable improvement.

Burr was not quite the catch the directors had anticipated. Henry believed that, coming from the public service, Burr was not well suited to the demands of a private enterprise. Burr's shortcoming, though, proved to be his certifications of the percentage of copper in the ores sent to Port Adelaide. Independent assayers found these seriously 'optimistic'. This embarrassed the SAMA, and Henry as Secretary, when those ores were assayed on reaching London or Swansea and found to be wanting. Furthermore, Burr's reputedly wild personal life unsettled his workforce. It was alleged that he and his wife hit the bottle very hard. If this were not enough, Burr had an eye for the workmen's wives as well. Neither were their daughters considered safe, even in daylight, as John Adams reported he had heard it from Henry. Adams was stepfather to John Benjamin Graham, the Burra's largest shareholder who had returned to Britain early in 1848. Adams further reported to Graham, though, that the SAMA directors were not entirely as one on the latter vice; Captain William Allen was inclined to feel that the Superintendent should be allowed to 'get his greens where he [could]'. As an ex-captain with the East India Company's fleet, perhaps this director's sentiment was not too surprising, but it was ironic, really, that Allen was simultaneously helping to found the Collegiate School of St Peter, which was about to open its doors to the boys of 'respectable parents', as it was advertised. Henry's sons would be educated there.

A hand-drawn cartoon in one of Henry's scrapbooks graphically depicts a lubricious Burr, caricatured as a strawberry plant. Its leaves represent arms, the face is reminiscent of the classic image of Mephistopheles, and a shoot stemming from his lower parts has a large strawberry at its end. The caption reads: 'the large strawberry as grown at the Burra and cut by the Superintendent'. Maybe Burr had exceptional horticultural skills. Such a fruit would have been unusual, as Burra was an acknowledged desert in terms of private gardens. Dangling from the arm-like right leaf is a pair of scales. This appears an allusion to Burr's responsibilities in ore evaluation. They were turned into a weapon against him, however, by the cartoonist who balanced 'gentleman' in one pan against 'Dangerous, Inebriant, Proud, High, Life', on top of each other in the other pan. The significance of 'Life' is a matter of conjecture; undoubtedly, High and Life were meant to go together, an allusion to Burr's racy goings on. One imagines that this cartoon was circulated only among the gentlemen of the board and was certainly not for any lady's eyes.

A rather more decorous cartoon is a nice satirical depiction of the Burra area; it is either a play on this inland district's summer dryness, or it uses hyperbole regarding a wet season, which the southern spring of 1848 was. It portrays a fjord-like vista down a gorge between the Burra hills with a couple of three-masted barques at anchor in the distance! Satire on Burr seems the more prevalent mood, though, with another sketch, of an anthropomorphic wild animal of sorts, carrying the legend 'A live Boar as seen superintending Burra Burra Mines'.

Toasting Success

By the middle of 1848, disillusionment with Burr coincided with the Year of Revolutions on the European continent. The initial outbreak, in France, caused chaos in trade and industry in Britain. The plight of British workers on French rail projects, thrown out of work and their savings appropriated by the revolutionary government, was relayed to one of Henry's friends by a relative in Wales. The demand for copper fell, not surprisingly when it is considered that, for example, each railway engine contained about 10 tons of it. News of the fall in the value of copper ore, the world price of which was set in Swansea – 'Copperopolis' – reached Adelaide by the April.

Up to this point, all had been going swimmingly for the Burra shareholders. The price of their £5 shares had risen astronomically, to a peak of over £200, with juicy dividends to match. J.B. Graham, as largest shareholder, was deriving an annual income of at least £16,000 from his investment in 400 shares. Stepfather Adams teased him with the promise of a large increase in his earnings the following year. In the current year, he expected Graham would gain *only* £20,000.

As the revolutionary outbreaks spread through mainland Europe the share values fell and the directors looked to cost cutting. The men would have to take lower wages and Mr Superintendent Burr, on his fat £700 salary, locked in for three years, would have to go. The over-optimistic assays were the pretext, and one wonders whether accusations of drunkenness and philandering were exaggerated as a fallback for the directors. Dr von Sommer had had his lawyer out-face similar accusations – nothing to do with the SAMA – against himself in the press the previous year. Acting for him, R.D. Hanson made it clear that, while his client had been found guilty of drunkenness, he had been acquitted of wishing to 'deflower an innkeeper's daughter and servant'. In Burr's case, though, Ayers had warned the Superintendent in the August that, when the directors visited, he would want a word with him about matters other than those covered in the official correspondence. It is ironic that the *Illustrated London News* later in the year credited none other than Superintendent Burr as being the one who had cleared 'bad characters' out of the Burra mining community.

Luckily for the SAMA, the right man to hold the fort was already in its employ. Another Cornish migrant, Captain Henry Roach served as Chief Captain under Burr with distinction enough for the board to advance him to the overall on-site management of the mining operations. Ayers' contemporaries would not have recognized the term 'downsizing' but they would certainly have recognized the concept. Roach could be relied on to continue his activities as Chief Captain *and* take on the duties of Superintendent, if without the title, on his modest £300 salary. Nothing could have pleased the board more, except perhaps a sudden steep rise in the value of their ore again, than to have the combined management roles at a fraction of the former cost.

To cement Roach's position in the organization, he was given both a private party and a dinner, attended by the Ayers. The party was held at Prospect House, the luxury residence of J.B. Graham. In Graham's absence, his stepfather, John Adams, had the run of the grand house. Adams recounted the evening's activities in the journal he kept as a basis for his letters to J.B. Graham. The party had a homely, in-house feel, with even Henry Ayers' city-office staff, MacDonald and Brodie, whom he considered 'all good and true', invited. At table, Mrs Adams, Graham's mother, took the Chair, with Henry Roach on her right and Henry 'Ayres', as Adams often spelled it, on her left. The spelling of his surname was obviously a touchy point with Henry; even the newspaper notice of his appointment as Secretary had got it wrong. Henry set his mine officers straight: '"Ayers" is the correct way.' It is curious, though, that on one occasion, in the index of one of his scrapbooks, he slipped, spelling his own name in the Ayres manner. At the party, there was singing, dancing and little acted scenes, with Adams confessing that he had never enjoyed such performances more since he 'became a playgoer'. They were in a 'continual roar of laughter the whole of the evening', all agreeing that it was the most enjoyable night of their lives – 'Ayers was never more at home as well as Capt. Roach and all parties'. The Adams looked on benignly as the younger element caroused but were not found wanting during the singing; Adams performed his 'Bluebells' and his wife sang her 'Violets'. Although Adams does not report it on that occasion, Henry may have sung 'Have you not heard of the Monster Mine?' As Ian Auhl (1986) tells us, this was Ayers' favourite party-piece. Henry pasted two copies of the text in his own handwriting into a scrapbook:

> Have you not heard of the Monster Mine?
> There's devil a man to be got to dine,
> There's devil a clerk who will pen a line,
> At my behest or thine.
> They are all gone forth to the homeless North
> To gaze on the Monster Mine.
>
> [J.W. Macdonald, Acting Colonial Treasurer
> – later Stipendiary Magistrate in Kooringa]

Most of the revellers drifted off by 1 a.m. and, with the house quiet, Adams, Roach and Ayers kept the fire warm, talking of the mine into the small hours. It is interesting that it was Adams, who saw himself as proxy for his shareholder stepson, together with Roach, the new overall mine captain, and Ayers, the Secretary, who chewed over the mine's future development, while the directors were unrepresented; it is perhaps an indication of a more managerial role, which Henry was steadily assuming. The Ayers were staying the night anyway, and in the morning, following a hearty breakfast, they were carried back to town in the Adams' carriage.

What sort of a night they had passed, though, is questionable as a ferocious storm had blown up. Notwithstanding Adams' best efforts, a blind from the main house and part of the verandah of a cottage in the grounds were carried away. In the aftermath, Adams fell down on his Christian duty the following morning. Despite a personal invitation from the newly enthroned Bishop, he was too unwell to attend the consecration of Walkerville Church, to the city's north-east. Adams assuaged his conscience with a five-guinea donation.

Roach's more public 'do', a couple of nights later, was a dinner given by the directors at the Freemasons' Tavern in Pirie Street. Adams was not there, but noted that Director John Waterhouse was missing, still feeling the effects of yet another party they had all attended, the week before, at Prospect House, on 'declaring Dividend Day', as Adams called it.

That, too, had been a wild night, in Adelaide terms, with local musician Bennett providing entertainment. The party-goers arrived merrily shouting 'sharp 6' as Adams had written upon the invitations. When Henry Ayers' health was being drunk 'with all the usual encomiums', Adams brought the excited crowd to a sudden silence; he said he could not let the moment pass without bringing a serious charge against the Secretary. He played it straight, with all desperate to know what dire accusation might follow and, despite the chill of a winter's evening, Henry must have broken out in a sweat. The 'crime' turned out to be that Adams had called at the Ayers' house, one Sunday, and found Henry surrounded with business papers and books instead of observing a day of rest. It cannot be said to be much of a joke, retold for today's reader. In a fairly devout colony, though, with Adams the committed churchman, it made its point and the company was uproariously amused – at least, Adams told his stepson that they were – and 'all ended extremely well'. Some overdid the drinking and John Waterhouse, particularly, put the fear of God in the others by saying he would 'knock every man down in the room'. He was packed off home in Jamie Chambers' trap, under the eye of Charles Beck. The Chairman, however, insisted on riding outside, afraid of what Waterhouse might otherwise do to him. As Adams informed Graham, Waterhouse, by his own admission, had 'never come to himself since he came to our party' – and he, a good Methodist! The Adams went to town, taking overnight guests, the Ayers, with them. Before the Mesdames Ayers and Adams went for a spot of shopping in Hindley Street, a 'snack of sausages' was enjoyed by them all at the Ayers' house. On other occasions, after attending a function in town, the Adams had their hospitality returned. After February 1849, the Ayers could conveniently put Mr and Mrs Adams up in their 'New Bedroom'.

Three years of the Burra Mine had seen it become the most prominent enterprise in the South Australian economy. It immediately supported a thriving mining town of almost 5,000 people and served as a strong catalyst for advancing the colony in which the Ayers had cast their lot. Henry was establishing himself more and more at the centre of the SAMA's activities, and of its social world, embraced by the likes of its suddenly wealthy directors and shareholders. The importance of the wealthiest – J.B. Graham – to him, and of Henry to Graham, was about to grow, shaping both their lives.

3

FRIENDS WHEN AND WHERE YOU NEED THEM

John Benjamin Graham's blood must have run cold, thinking about his dividends, when a whisper reached him in London of rumpus at the mine, added to the problems with Burr and falling ore and share prices with Europe in ferment. Reassured that it was just a rumour, he must have drawn a huge sigh of relief because the previous few months had seen considerable expenditure.

His visit to Britain had several purposes. He could visit the SAMA's various agents and associates, and repay debts of honour that he felt on account of his late father's bankruptcy. His third errand was more personal. As a 35-year-old bachelor, and recognized by *The Times* editor no less as the 'lucky emigrant', Graham was seeking a wife.

There may also have been an unspoken, fourth reason for the trip, resulting from a strange anonymous letter he had received later in 1847. Signed 'Eureka', the letter ardently urges secrecy, citing Graham's recent initiation into the brotherhood of Freemasonry, Graham's 'Passport to every grade of society throughout the privileged world'. The writer urges Graham not to sell out prematurely, his share values rocketing even as Eureka writes. He is also urged to hurry to Europe to broaden his mind, and return fully rounded, ready to endow charitable institutions, which will perpetuate his name.

Whether or not prompted by this letter, while visiting the Swansea smelting sphere in August 1848, Graham dined with the 'leading philosifer of England' – probably Michael Faraday – at the convention of the British Society for the

Advancement of Science. He also visited 'Mr. Nevel', the proprietor of a sizeable copper smelter at Llanelli. Nevill's nearby competitor, the 'Spitty' works, was in process of seeking direct access to the Burra's ores, bypassing purchase by auction at the Swansea 'ticketings'. Graham confided to his journal that one of Nevill's daughters had no liking for music, 'so she would <u>not</u> suit me'.

Another unsuccessful South Wales candidate for Graham's affections was Jane Coke, sister of South Australian immigrant James Charles Coke. Jane told her Adelaide brother that the family was 'agreeably disappointed' with the Burra magnate. Amused by Graham's conceit about his singing voice, she admitted they had to make allowances for his lack of education – perhaps as obvious to the Cokes as to 'Eureka'. The magnate was more than generous, though, in presenting his new friends – or possibly their church – with a powerful organ, costing £1,000. Jane conceded that her lack of real keyboard proficiency had made a closer relationship with the wealthy visitor impossible – 'we were good friends and as such we parted', she joked. Besides, for heirs to his brand new fortune, Graham would need to look to someone younger than a Jane Coke already into her forties.

Back in London, Graham soon found a girl meeting his requirements. He took Louisa Rymill to choose a piano for his new house in Brompton, Vere Lodge. Costing £165, it did have a 'most splendid tone'. The Lodge's 15 rooms they decorated lavishly, including with artworks bought at Stowe House, home of the Dukes of Buckingham. He also spent an agreeable evening with the celebrated artist Landseer, of 'Monarch of the Glen' fame, and probably purchased a work from him. Add to the list a diamond ring for Louisa, and a coach for £1,200, second-hand after the Duke of Buckingham, and expenditure was mounting. When news of the Duke's vehicle reached Prospect House, Adams salivated over the 'commotion' it would cause when Graham brought it back to Adelaide.

There was the rub – would he return? Louisa would not leave her family. Graham reported her 'all in a fidget' about living in South Australia, as well she might have been; the whole romance had been a whirlwind. Graham would therefore have to settle down in married bliss on that side of the world, the wedding taking place in March 1849.

Trouble at the Mine

By February 1849, the bombshell of Graham's non-return struck his Adelaide circle and Henry Ayers therefore penned a newsy letter. The rumour of trouble at the mine was true after all, and was nothing less than an all-out strike. He told Graham that the workers on 'tribute', the men who contracted in 'pares' or 'pairs' to work a certain lode for a percentage of the value of the ore they raised, had refused to bargain for 'pitches' at the October 1848 lettings. It was in retaliation against the SAMA after the directors declined to settle their grievances over their losses through Burr's incompetent assays. The rewards on offer from the cost-cutting, and punitive, board were simply below what they could stomach.

The board's September carriage jaunt to Burra had begun as usual, with Ayers and the deputation of directors setting out light-heartedly. Adams and Graham's mother had saluted the carriage, she with a 'White Flag' – doubtless her handkerchief – from a window, and he waving his hat from the garden as they passed along the wall around Prospect House. Jocularly, the travellers rose in their seats to return the salutations and the visitation began in the confident, easy-going spirit with which the week had begun, when the new Governor and his wife had honoured the SAMA office with an official visit.

From his physiognomy, Adams concluded that Governor Young would not be short of something 'in his noddle'. Lady Young had her accomplishments, too. A fine horsewoman, with the agility of her mere 20 years, she saved herself from a nasty fall by turning a 'Somerset' – and that while in the early stages of pregnancy. The couple's greatest achievement, though, was to go about the town 'from shop to shop as if they were nobody', making themselves 'very familiar'. From the point of view of mining shareholders, Governor Young had an extra appeal, in removing the royalty charged on ore exports, a policy of his predecessor that had generated the colony's first rumblings of 'independence' sentiments from the Mother Country, and these in a colony barely out of its first decade.

The new gubernatorial couple were received in the SAMA's office by Secretary Ayers and two directors. This was not the £10-a-year, one-room affair at Smart and Bayne's. Henry now operated from an administrative centre prominently positioned on the corner of King William and Rundle Streets, at the heart of commercial Adelaide. With Governor and Lady Young visiting, John Adams obviously could not keep his nose out and, on his way home, he happened upon Captain Roach at the ford over the Torrens, on his way back to Burra. Everything in the SAMA garden must have seemed lovely, its premises hallowed by an official visit, its new Captain-in-charge returning to take up control, and now a 'holiday' to follow for Ayers and the directors in their hotel, in their mining town.

It turned out no holiday. The men were striking and 'taking possession of the mine', with Henry and the visiting directors besieged in the Burra Hotel. The editor of the *Register* joked that they were protected from the vehemence of their predominantly Cornish workforce due to the name of its landlord,

Wren. As the old Cornish proverb went: 'Neither strike robin nor wren, be you boy, maid, or man.'

In town, meanwhile, the Secretary's 'good and true' understudy, MacDonald, approached Commissioner Dashwood to consider sending up any police he could spare to quell the 'disturbances'. There was not really much unrest, but the board was spooked. Director William Paxton, a Rundle Street chemist, rode with the Commissioner and about 20 officers, according to MacDonald, who also warned Henry that he might be in hot water from the other directors. They were 'surprised' that he had said nothing on the subject of 'Mr B'. Burr actually got his marching orders during this visit.

The well-armed mounted-police posse was, in Adams' words, 'expecting warm work'. After busily putting names on pews in St John's church, Adams called to see if Anne Ayers had any up-to-the-minute news from the mine. She was quite heavily pregnant and Adams found her 'very poorly'. There was no news but she was 'all in suspense fearing a collision'. With Henry away and possibly in danger, Adams could do no other than take her back to Prospect House and grandmotherly Mrs Adams.

The mine workers had written, with due deference, appealing to their masters over the valuation of the ores, which were consistently considered of lower quality on reaching town than previously agreed with Burr. This was no trick by the directors, of course, who had been themselves embarrassed when ores reaching Swansea were found to be below the stated quality.

The board scorned two written appeals from the united workers – get back to work and we will talk to one or two of you 'uncombined'. The men scrambled together their individual tribute heaps so that no one could settle with the bosses unless they all did. Impasse. Sympathy action by the draymen lasted a mere three weeks. The miners, once the grievance was settled – on the board's terms – were kept from working, and thus earning, for two months, until New Year 1849, as a kind of lockout or cooling off period. The strike was actually mostly carried off in good humour; two blacklegs, hauled out of the mine, were tied back-to-back and carried shoulder-high on a handbarrow all the way to the Burra Hotel, suffering nothing more than the laughter of onlookers.

An arrival of new migrants allowed Henry to insist that Roach maintain the low-pay regime: 'What fools you must be to pay ore-dressers thirty shillings a week. If you cannot procure men at twenty-one shillings, I will send you as many as you require, for there are hundreds of men starving in Adelaide.' Although a handful of strikers was let off by the local magistrate with light punishment, a couple of dozen ringleaders were rooted out and their names blacklisted in other South Australian mines.

Despite Henry's hard-headed business line, he did privately respect the tenacity of the striking miners. He wrote to Graham that they had shown more spirit, or deeper pockets, than the draymen. He also believed that, though shareholders like Graham – come to that, like himself – might have forfeited a dividend, their sacrifice was well worth the increased power the management had gained over its workers. This he regarded as 'priceless'.

In more humorous vein, Henry told Graham that Charles Beck, the SAMA Chairman, had said he would hate Frenchmen for as long as he lived for devaluing the ore! For more than a moment, it looked like Britain would be caught up in revolution, too. Worried readers of the *Illustrated London News* were provided with the words and music of 'Peace at Home be our Prayer'. A last attempt by the Chartists to pursue their ten-year-old cause, by holding a mass demonstration on Kennington Common, was defeated as much by the weather as by the Home Secretary's special constables.

Charles Beck was not alone in mourning his losses; all who had purchased ore were in the same position. Henry also had a stern message for Graham, who was inclined to be too free with his inside knowledge about Burra; he must keep his mouth shut in London. Despite all the problems, 'Burra is still Burra and will yet do better than ever'. By then, Henry could afford to view the upheavals with some perspective. Things had turned out reasonably well, despite 'French Revolutions and Burra Burra D[itt]o'. Adams, with more immediacy, recorded the excitement of the September return journey from the mine; the directors' coach overturned almost immediately upon pulling away – no doubt to the delight of the defeated miners. Police Commissioner Dashwood, sitting with Ayers and Chambers on the box, had made a panicked grab for the reins when he feared the horses bolting, probably making matters worse. Beck had to remain behind, Ellis had an injured leg and Kingston a cut head. One imagines Henry to have sustained more damage to his pride than anything, having been thrown clear, but into a thorn bush.

A Novel Proposal

Henry replied to an official approach to the SAMA by one Gregory Seale Walters. Arriving in the colony in July 1848, Walters sized up the best means of promoting the interests of the company he represented, Schneiders' Patent Copper Company (PCC), of London. H.W. Schneider had just expanded the family's interests into copper, developing the Spitty smelting works west of Swansea. Graham visited these works, noting that they were run by the very company, and on the same Napier's Patent, as was about to set up operations in South Australia.

The Schneiders had observed the rich South Australian ores reaching Swansea and decided to pursue the possibility of setting up a fully fledged copper smelter at source. They could then avoid the pricing competition of the Swansea ticketing sales. The South Australian mine companies were unlikely to complain as it would remove the cost of shipping to Britain – a risky business, over such a distance, especially at a time of crisis such as this Year of Revolutions, when the price in Swansea could be much lower, on arrival, than anticipated when freight, insurance and credit had been arranged some four or five months before.

There was, further, the chance of losing an entire cargo, as when the *Brechin Castle* foundered near the Mumbles Head, within sight of the port of Swansea, during a February gale in 1847. Crew, passengers and cargo were all lost. Though the insurance claim was paid promptly, Henry had to write to complain of the cut taken by the agents. How much better, then, for miner and smelter, to eradicate the distance between their operations. Both could then serve the high-demand Indian and Chinese brass markets virtually on South Australia's doorstep, thus stealing a march on Swansea.

The field was not by any means open for the PCC to simply walk in: although the SAMA's own smelting forays had borne little fruit, possibilities in the other Australian colonies or locally were being explored. For example, Charles Penny and William Owen planned a smelter at Apoinga, 20 miles south of Burra, and offered to smelt the entire Burra output. Cautious Henry was circumspect, suggesting they deal initially with a sample 5,000 tons. Within days, the last nail was knocked into the coffin of the SAMA's own smelter, with the Dreyers' dismissal. Henry had concluded that even the richest quality ores would never pay using their methods, though the Dreyers had not, of course, been given the opportunity to smelt the rich ores, only poor ones not worth shipping.

David Bannear (1990) asserts that the Patent Copper Company, the PCC, came to South Australia through Ayers' lobbying. This is difficult to substantiate through the official correspondence. *The South Australian Gazette and Mining Journal* did make reference to possible British-built smelting works at both Burra and Port Adelaide. This grandiose plan was possibly the Schneiders', and, if so, it would take about 14 years to come fully to fruition.

London merchant Frederick Huth informed Graham on his arrival in June that a ship would bring out an entire smelting establishment. Huth's son-in-law, Gregory Seale Walters, was to have overall management of the PCC's venture, which was probably not limited to a possible Burra connection. In the good two months G.S. Walters was in the colony ahead of the barque *Richardson*'s arrival with the smelting requisites, he had the opportunity of considering the potential of each of the colony's mines before choosing the most advantageous site for operations. That a decision on Walters' part was still pending by early September is clear from Henry's reply to him of the eighth of that month. Henry's tone is one of pleasure at the prospect of the two companies working for mutual advantage, and the whole thing appears to be coming as an unlooked-for windfall. Further, he stresses to Walters that the Burra mine has plenty of 10 to 12 per cent ores 'at grass' (the surface) – not worth sending overseas. The PCC's immediate role, then, was in mopping up this otherwise fairly valueless produce. The PCC had going for it the right to use the Napier Patent, which enabled the numerous processes of the standard 'Welsh Method' of copper smelting to be reduced, controversially, to as few as two. When the principles of an agreement between miner and smelter came to be thrashed out, Henry and the SAMA insisted upon a right to inspect the PCC's activities, at will, to check that it was using the Napier method.

Henry wisely suggested a trial. He had in mind an auction of ores at the Port – a kind of mini version of the Swansea ticketings, though he said he would not necessarily adhere exactly to those methods, which he felt were not entirely advantageous to the miner. Robert Toomey considers that the ticketing system was in fact regarded as fair for both mining and smelting interests. The Swansea merchants and shipowners Leach, Richardson and Company put the case as strongly as to claim that a miner did better out of the Swansea procedure than he could in London or Liverpool. They took pride in the fact that the assaying was both repeated and public. By later 1848, Henry might well have been sceptical of such local claims especially as the cartel of the 'Smelters' League' was manoeuvring to minimize the impact on their trade caused by the Year of Revolutions. When news of the August sales that Graham witnessed reached Adelaide, by the end of the year, it was clear that the value of Burra ores sold had slipped another 4 per cent.

A location for Walters' establishment was also canvassed in his negotiations with Henry. If near the Port, the SAMA would be responsible for carrying ores there, but Henry proposed another option: land near the mine could be made available to the PCC for leasing. This novel approach had possible pitfalls for Walters. As a far greater weight of fuel was required than ore to produce a quantity of pure copper, it had always been the practice to take the ore to the fuel. Bunce and Davy were, for example, setting up a smelting venture near the Port where a coal supply, though expensive, would not cost much in carriage once landed, should they need it if locally produced charcoal proved inadequate.

Henry's proposed 'short-straw' option in fact had some appeal for the canny Walters. He had seen what the SAMA had not: traipsing the almost hundred-mile bullock track from Burra to Port Adelaide was not the smartest idea. His fresh eye saw a much shorter way to the coast, striking out in a more westerly direction. True, there was no road, but the more southerly route in use hardly merited the description anyway, and next to none of the westerly land was taken by pastoral or farming interests. All that was required was a decent landing place for goods, in and out, at the head of Gulf St Vincent.

Henry proposed a face-to-face discussion, suggesting Walters meet him together with Chairman Charles Beck. This is a strong indication of the sort of power Henry was starting to wield within the Association, born of his detailed knowledge of the everyday intricacies of the business which occupied him full time.

Almost immediately, though, Henry was off on that ill-fated visit to the mine and would not return to town, in his thornbush-damaged state, until 21 September. Not that Mr Walters was entirely out of his hair during the Burra crisis; for no accountable reason, other than the fact that he happened to be on the spot, Walters tried his own managerial skills upon the striking miners, despite having allegedly received some 'maltreatment' from them. According to his own letter to the *Register* under the pseudonym 'An Eye-Witness', he stated how he had tried to assure the men that the SAMA directors harboured nothing but warm feelings for them, and he was sure that all outstanding grievances could be resolved. He sided with the directors, though, regarding not giving way before mob coercion. Although not on individual contracts as such, the men were contracted by 'pairs' and, as each pair had grievances differing in detail, the directors would only consider them severally. The SAMA hoped to divide and conquer their disgruntled workers. How embarrassing, though, for Henry and Beck to have to commence negotiations with this London businessman while activities at the mine were at an enforced hiatus.

An Insider's Knowledge

Henry would find Walters a very tricky, if disarmingly pleasant, customer. Henry may not have been entirely in the dark, though, about the PCC manager's recent history as he had someone in his immediate circle in a position to know Walters' full background. This was James Coke. While the Bank of South Australia's local manager, Edward Stephens, also knew to be careful of Walters, it is unlikely that he would have passed this information on to the directors or Henry; the SAMA banked with his competitor, the Bank of Australasia. As Stephens' institution was cautiously helping to fund the PCC's venture, it would not necessarily have been in Stephens' interest to let Adelaide know what he knew. He had been warned by his London boss to be wary of Walters, described to him as 'apt at financial schemes and rather plausible'.

Coke, whose insufficiently musical sister Jane had entertained Graham in Wales, was involved in merchanting, linking two parts of the world then so wrapped up in the copper trade. The Ayers and Cokes had emigrated on the same ship, the former in Steerage and the Cokes in the Cabin. Whether they met on the voyage is unknown, but both Adams and Graham regarded Henry and James as friends. James was well placed to fill the gaps in Henry's knowledge in this new sphere of activity. Coke's brother-in-law, the well-credentialed Captain William Llewellyn Powell – who had conveyed the visiting Czar back to Russia in 1844 – maintained a faithful correspondence during the years of James' absence in the colony. He supplied James with both merchandise for selling in Adelaide and inside information on James' parent merchanting house in Swansea: Leach, Richardson and Company. These merchants believed James had some influence within the SAMA and counselled caution: the SAMA should not constantly send its highest-quality ores to the Swansea market – at one point, Dr Davy had assayed some 'red (horse flesh)' ore for the SAMA at 77 per cent metal. Captain Powell could give his brother-in-law information on ore assaying, gained from an employee of a principal Swansea smelter, the Vivians', augmented by what the Captain himself had gleaned from books. Clearly, a well-briefed James could position himself to best advantage for his own possible involvement in the copper trade, being a valuable contact for his friend Henry Ayers.

In early May of 1847, the Leach, Richardson brig *Appleton* was to sail from Port Adelaide, carrying valuable cargo consigned to Coke and associates. She had aboard 335 tons of malachite ore from Burra and the celebrated explorer Captain Sturt and his family.

Although James was apparently a scrutineer of ballots for the 1847 election of directors, he appears to have risked putting himself offside with the SAMA just prior to the *Appleton*'s sailing. He made bold to enquire of Ayers the value of the various quality ores being shipped. Henry replied rather tersely that commercial confidentiality prevented him from giving out such information. However, within a couple of days, he was personally communicating with Leach, Richardson about their terms for handling SAMA ore consignments as conveyed through James. Henry rode one of his hobby-horses, begging them to ensure that the Burra produce was not contaminated by ores of any other mines as it was of particularly good quality.

An official connection between Henry and James is next documented in April 1848 in the aftermath of a nasty bout of illness. Henry had been ill for a fortnight with a threatened inflammation of the brain. It had not been a trivial event, then, and must have caused quite some concern in the Ayers household. Once back in his normal rude health, Henry was in touch again with James about a tender to carry ore to Swansea now that the *Appleton* had made her second voyage to Port Adelaide. Henry wrote a long letter to Leach, Richardson, concerned that the *Appleton*'s previous cargo had been sold privately and not through the ticketings. He made it clear that, in future, its ores should be sold on the open market.

For their part, Leach, Richardson also had their strictures to put on him: they were not prepared to accept advances on ore shipped, unless on vessels equipped with 'platforms and trunk[s]' such as the *Appleton*. James had passed on a letter to Henry stating that they were happy to take ore as ballast on ships not so fitted, but if it was an out-and-out ore cargo, their requirement stood. Henry betrayed his naivety in the copper trade here, admitting his unfamiliarity with these fitments. It is probably an indication of the ad hoc nature of the shipping available to carry South Australia's copper ores up until this point, because Swansea ore-carriers, used to plying to Cuba and Chile, had generally been equipped with these fittings for more than a decade; they enabled the ore to be loaded in such a way as to improve the vessel's stability and handling qualities. The Leach, Richardson writer informed James, in a letter of early 1849, that the *Appleton* had weathered a very bad storm on its last voyage. The writer put the ship's survival down entirely to the good order in which she was kept and the fact that there were 'no corners cut'.

After the excitements of September 1848 at the mine, and with plenty of ore to ship, Henry next accepted James' sole tender to ship per the *Emperor of China* – one of the largest vessels to ply out of Swansea in those days. James was soon at Prospect House himself, bringing Captain English of the *Richardson* with him. The greatly anticipated arrival of his barque had taken place on 3 October. It must have been a source of considerable apprehension for the smeltermen and their families to find the fabled 'Monster Mine' suffering a crippling strike and continuing out of action.

Of Gregory Walters, James had information: Walters was being given the opportunity to try to rebuild his reputation, damaged when his London firm went bankrupt. Henry wrote to J.B. Graham: 'the tales that could be told!' In Henry's scrapbooks a newspaper obituary at the time of Walters' death states that the man had had to unfairly shoulder much of the blame for the collapse of A.A. Gower, Nephews & Co. His perceived business acumen must have held some fears for the SAMA Board, because another of the pencilled cartoons in Henry's scrapbooks shows a portly Walters, at the head of the directors' table, swallowing the pygmy figures of the directors, who are sliding inexorably down an inclined tray, or trencher, into Walters' gaping mouth. The sketch is entitled 'Walters Vivat' and beneath the picture is the subtitle 'Greased Wheels' – which might have been a reference to the London manager of the Bank of South Australia, E.J. Wheeler. The cartoon also bears the elliptical legend: 'No wonder that so short a time the trick "Ala" [*sic*] Walters did in who had the Chairman [of the SAMA Board, Charles Beck] at his <u>beck</u>, and t'others at his <u>bidding</u>.'

It was credited to 'A holder of Burras' – that is, a holder of SAMA shares. Unlike the other cartoons, this one was apparently copied from *The Mercury* of April 1849.

Whether Henry drew, or copied, the cartoons himself, or gathered up the efforts of more artistic, or less attentive, board members, his collection of them bears out the existence of a wry sense of humour. This is also apparent in his correspondence, but finds no echo in descriptions of him in, for instance, Ian Auhl's (1986) *The Story of the 'Monster Mine'*.

If Gregory Walters had a chequered past, so too did his deputy-designate, Thomas H. Williams. Coke learned that Williams, following his own business failure, had been found a post at the Spitty works. That proved the spring-board for lucky Williams to obtain the Superintendence of the PCC's South Australian works-to-be, doubling his salary, at £500. Rumour will spread in the business world and in no time at all Captain Powell was writing to beg Coke to confirm that Williams was not getting an 'absurd' £2,000 per annum for this colonial project! Henry Ayers was well in a position to know, then, that the management team of the PCC was not necessarily as competent as, certainly, the editor of the *Register* chose to portray it.

The St John's Circle

Almost the entire cast of players in this colonial copper-industry drama came together, weekly, as members of the congregation of St John's Church, in the south-east quarter of Adelaide. This place of Anglican worship was affection-ately known as 'St. John's in the Wilderness' due to the lack of habitations between it and King William Street. One or two of the copper-industry cast met each other 'on the square' as Freemasons, some others as Oddfellows, but St John's united nearly all of them.

In his 1848 letters to Graham, John Adams made frequent references to Ayers family attendances at St John's. The Adams might pick up the Ayers en route to a service, or call back at the Ayers' house afterwards. Adams, a very committed churchman, was obviously encouraging Mrs Adams to become more devout. By the October, he could report to Graham '. . . your Mother, who is now quite the Church woman' almost in the same breath as jesting about her gad-about ways: 'Your dear mother getting quite a rake – I cannot keep her at home.' He made free to contribute on the absent Graham's behalf, not limiting himself either to new Anglican church-buildings.

Maybe Gregory Walters and Thomas Williams saw which way the wind blew, or perhaps they were just following their own custom when, by early February 1849, Adams reported that both had rented pews and attended services regularly. He had even succeeded in persuading Ayers to rent a pew. It was really like the Return of the Prodigal, for Adams, who was relieved to be able to tell Graham that Henry Ayers was showing more signs of Adams' sort of piety. Adams had been concerned that Henry had shown an 'indifference in preparing for Eternity – I have reason to think that they are beginning to see the reasonableness of my hints'. Adams' mock reprimand to Henry for being caught working on the Lord's Day had been couched as a joke, but obviously had had serious intent. Observance of the Sabbath had been amusingly topical with a report in the *Register* of a shoe-repairer's setting about soling a boot on a Sunday morning. The boot's owner had made a temporary repair by lining it with a religious tract. When the repairer winkled out the stop-gap, he dropped his work and fled to the nearest church. The tract was entitled: 'Remember the Sabbath Day to keep it Holy'.

Adams, at 53, certainly had his eye on his own Eternity, devoting much time and effort to the governance of the new Pulteney Street School, delighting in the children's shows of piety at a service at Holy Trinity. His reputation for good works would rise further when, early in 1849, he hoisted a huge Union flag over Prospect House – 'Graham's Castle' as he called it – and hosted an enormous fête for all the children of the Sunday Schools of Holy Trinity and St John's. Henry brought Anne to the fête, on his way north to Burra, with the children and a nursemaid. Adams was most concerned to be perceived to have left good works behind when 'mouldering in his grave' though, on a more worldly note, he also admitted to trying to build up good will in the colony for J.B. Graham. Graham would be back, Adams was confident, once he had had a 'regular surfeit of this world's Bubbles and Blow Bladders'.

Adams also noted that Walters would arrive at St John's in his own carriage – the exact same model as the Adams' own, though Adams had lost his nerve for driving personally; he preferred his driver to be on the box. His perception of his own dignity might have had something to do with it as well because he went on to refer to being 'the Father in law' – actually, stepfather – 'of the great Mining King'. The ripples from the wealth of the Burra were spreading even more widely because Mr Secretary Ayers and family visited Prospect House in late 1848 in *their* own carriage. The lad from the Portsmouth law office had certainly come up in the world in a mere eight years, and that without Graham's 'passport' of Freemasonry. 'Poney' had given way to horse, which could now join in pulling a carriage.

The family had grown, too, with the addition of the Ayers' first daughter. Frank and Harry had been followed by Frederic, in March 1847, but daughter Margaret Elizabeth became the apple of her father's eye. Baby Maggie was an overnight guest, together with her parents, at the Adams' on a number of occasions, proving herself 'quiet and very good'.

It is very difficult to assess just how close Henry was to the SAMA's biggest shareholder, J.B. Graham, prior to Graham's departure for Britain but their relationship was certainly developing soon afterwards, not only through their correspondence, but through close links with the Adams. A mere month after Graham sailed, Henry wrote that he was visiting the Adams weekly at Prospect House. He may have been pushing himself beyond the call of duty because he even visited during his recent serious illness. The wealthy absentee obviously felt he could rely upon Henry, and on his discretion, perhaps a little more than Henry felt he could rely on Graham's. However, the foundations of a relationship transcending the bounds of business were being laid which would continue, with ups and downs, until 1870 at least. The letters that Henry wrote almost monthly throughout that lengthy period have largely survived in the Graham Papers and reveal much about his family and business circumstances through the years.

It is clear that J.B. Graham put more faith in Henry's management of money matters than his stepfather's because he put the purse-strings of his own South Australian finances, at least partially, in Henry's hands from the start. Henry, in a letter of 1849, diplomatically checked with Graham about the acceptability of Adams' spending. Adams, in addition to giving to this church or that, was also doing his best to improve the environs of Prospect House, sparing neither expense nor effort against drying winds and grasshoppers in planting the garden with vines, exotic plants from India and even some sent over from Britain by Graham. S.T. Gill's 1850 watercolour of the house and immediate garden gives a very clear feeling of their exposed position, on gently rising ground, looking towards Port Adelaide. Perhaps it is a stubby Mr Adams who strolls away from us up the side path, in his pale, summer-weight jacket, half-obscured from our view by his umbrella.

Friendship Strained

Henry's friendship with James Coke may have suffered some strain towards the Christmas of 1848 because Henry, in a very disgruntled manner, put in writing what he perceived as James' misleading him about the course of the ore-carrying *Emperor of China*. Henry's concern for the purity of his precious Burra ores was raised when he gathered she had put in at Swan River. This was against his agreement with James, entailing a direct voyage to Swansea. So agitated was he that he even rehearsed his aggravation while writing to the Cornish firm of John Williams, which really had no involvement in the matter at all, recounting that, if the cargo were contaminated because of this detour, he would see that both James and the *Emperor*'s owner would pay for it. For his part, James could only protest that he was not aware that the vessel would put in anywhere other than the Cape, for water. Eventually, this storm in a tea-cup passed and the Ayers and Cokes remained friendly.

For another of Henry's associates, though, there was trouble. The solicitor Bayne's wife twice approached Adams for a loan from Graham's funds, to help her husband avoid bankruptcy. Adams used the excuse that Graham had told him on no account to lend out funds during Graham's absence. In any case, it was too late for Bayne, who fled to escape almost certain transportation to one of the convict colonies for forgery and other misdemeanours. Though Smart and Bayne had provided Henry Ayers with his entrée to his SAMA position, he must have been very relieved that his name was no longer associated with that of Bayne.

4

THE YEAR OF EVOLUTIONS: TAKING HOLD OF THE REINS

From late 1848 until well into 1849 was a very difficult time for the SAMA directors and perhaps even more so for Henry Ayers, who was up to his eyes in it. The falling price of ore, Burr's sacking and the strike all coincided with the negotiations with the wily Gregory Walters over the establishment of a sizeable smelting facility. In the middle of this time of crisis, the smelting personnel had already turned up.

Burr did not go down without a fight, successfully represented in court by James Hurtle Fisher who won him his full contractual payout. Henry's own position within the SAMA became a matter of debate as, of course, it was his signature on Burr's dismissal letter. Was Ayers a 'partner' in the Association, and therefore liable financially if it lost the case? He was found to be simply an officer. Before it reached court, Burr had fun by urging on the striking miners, causing Adams to consider him a 'very foolish silly man – he will do himself no good'. Burr was abetted in town by John Stephens' editorial efforts in the *Register*, castigating the shopkeeping nobodies who had struck it lucky at Burra but were now 'grinding' the men, at the lowest possible wages. The nine directors filed suits against Stephens, for whom Adams had no sympathy: he had been 'writing of the Directors in his usual scandalo way'. Stephens organized his own benefit night in aid of 'freedom of the press'. The labouring classes turned up in numbers, due to free admittance, though only a paltry 20 shillings was collected. The directors' court actions fizzled out but Henry wrote to Stephens, on the decision of the board, pettily withdrawing the SAMA's advertising business from the *Register*. At least the miners themselves collected £50 for their journalist-champion.

Well before all this, Gregory Walters had explored his proposed dray-road to the head of Gulf St Vincent and Henry instructed Roach to load drays for the novel route. Walters had taken up the SAMA's offer of a 40-acre site, at a peppercorn rent, 'across the Creek from our [the SAMA's] Smelting House' – the latter relic converted to the mine's storehouse. The PCC should only have the Burra's poor-grade ores and must not service any other mine without the SAMA's permission. The various clauses of the agreement were airily assented to by Walters, without his ever actually signing anything, and he steadily whittled away at the fine detail to the PCC's advantage.

The niggling aggravation upon Henry and the board of this protracted haggling finds an amusing illustration in another cartoon in Henry's scrapbooks. The SAMA bull attempts to lash out at a dog [Walters] snapping at its heels, the caption reading: 'Bull soliloquy – Ah you little varmint if I only get you within reach of my hind legs.'

Frustration with the slippery Walters would be the order of the day for Henry and the directors for some time but Henry was delighted with the energetic way in which Walters, Williams and their men set to work. Walters' initial reception at the mine was inauspicious. He was booed by the men during their post-strike lockout, for his 'Eye-Witness' letter to the newspaper. However, by December, there was a ceremony for laying the smelter's foundation stone. The new works were generally known as the Kooringa Copper Works, taking their name from the SAMA's own township, in which the creek-dwelling miners were still reluctant to live.

The Welsh smelters had to work through a wet spring and fierce summer. Yet, by January 1849, not only was a very handsome furnace-house erected but, amid the rubble of a building site, smelting experiments had already begun. The SAMA and PCC honeymoon was in full swing and Henry wrote to J.B. Graham that they were 'doing all that intelligence and money can do'. A race was on, against the Bunces' Yatala works and the Pennys' at Apoinga, to turn out a significant copper product. The ore-shippers and the Swansea smelters would probably become redundant within a year, as far as the SAMA was concerned, Henry boasted. He closed business with a Sydney smelter undertaking some small-scale work for the SAMA, confiding in James Coke he was quite indifferent as to whether the steamer *Juno* was kept on the Adelaide–Sydney run for ore-carrying. The strength of Henry's position to speak for the SAMA, at this time, is underlined by a query of his to new SAMA lawyer E.C. Gwynne as to whether the Association could sue or be sued simply in the name of its Secretary.

Incidentally, while things were quiet for the *Juno*, she embarked on an eventful five-day pleasure cruise to Port Lincoln. With numbers of local notables aboard, including the SAMA's Waterhouse and Paxton, a fight broke out between two 'worthy citizens'. The moderate, sober South Australia of the colonizing theorists showed itself up in a bad light and, with the crew already drunk, it was considered safest for the steamer to return to Port Adelaide and start afresh. The rowdier side of Adelaide resurfaces twice again, in Adams' journal, when he notes for Graham a fight between two gentlemen over a 'Young lady of fortune', and that R.R. Torrens had smashed George Stevenson in the nose with a 'short stick', for satirizing him in his paper, *The Mining Journal* – according to Adams, most people's sympathy had been with the 'smasher' rather than the 'smashee'.

Adams noted that more 'pitches' than ever had been let to the mining 'pairs', though 20 'troublemakers' had not been given lettings, despite begging. At least one 'committee man' was still appealing unsuccessfully for work three years later. Governor Young had donned 'the blue shirt' and spent four hours touring the underground workings, pleased with all he saw. Henry, too, unafraid of getting his hands dirty, made many descents into the mine over the years.

Twice in January 1849, the Ayers were up at Prospect House. In particularly hot weather, they stayed the night, the 'babby' providing entertainment for the out-of-sorts Adams couple. Adams presumed they had been 'baked out of the city'. Prospect House boasted an underground room and perhaps this cool sanctuary was too much of a temptation for the Ayers to resist.

By this time, the mine's workforce was getting back into harness and the smelting works were progressing. In March, Henry could write enthusiastically to Graham about the 'best of goodwill' between the two companies. The first ore was delivered to the new smelting works shortly before the SAMA's AGM in the April and, by then, too, Captain Roach had sampled some ore which again was more than 70 per cent copper. Things appeared to be turning the corner, especially now the copper price was recovering as revolutionary Europe started to settle down.

Amid the optimism came J.B. Graham's news that his awaited return was not to be. His bride standing firm, he would have to content himself as a non-resident shareholder, rather than a director, losing first-hand knowledge and a direct say in the control of his prized asset. Henry declared to Graham his worry that there would henceforth be a smaller board. With John Waterhouse going Home, John Ellis sold up because of losses in his wool interests, and with Stocks Jnr. on his last legs – Adams then considered him 'imbecile as a

Director' – the remaining directors were inclined to limit the board size. Henry did not relish this, perhaps feeling that it would be more difficult for him to accomplish any 'steering effect' upon a tighter-knit body.

Of course, Henry welcomed Graham's marriage, enthusing that his and Anne's nine-year union had been a 'blessed state'. One wonders, though, if it came back to him through James Coke that the latter's family in Wales were a bit miffed at receiving no notice of the Grahams' wedding. Captain Powell spoke of their disappointment, describing Graham as 'entre nous a bit of a snob'. Before long, a copy of Graham's journal, in Louisa's hand, arrived for the Adelaide family and friends to pore over; Henry teased her husband for making her 'the complete copy clerk'.

His friend's resettlement prompted a hint of homesickness and perhaps envy. Henry admitted that he would quite like to head a possible SAMA London Head Office. The comforts of Home must have seemed suddenly very attractive. At this stage, though, with developments in the SAMA's interests growing rapidly, even the prospect of a holiday visit was out of reach. Still, there were compensations. Henry wrote enthusiastically about the prospects for both Graham and himself, for he, too, had been increasing his – at first modest – SAMA shareholding.

Henry attended the official opening of the Kooringa works by Bishop Short. The Adelaide clergyman took as his 'text' the image of 'Latimer's candle', linking the lighting of the furnace with the burning of the Protestant martyrs, Ridley and Latimer, by Queen 'Bloody' Mary in the sixteenth century. The Bishop's little in-joke, referring to Latimer's famous speech as the faggots were being lit – 'Be of good cheer, Master Ridley . . . I trust that the candle we light in England today shall never be put out' – does not appear to have struck anyone as in doubtful taste. Henry certainly considered it 'a very appropriate address'. Henry described for Graham the sizeable furnace house which, at over 200 feet in length, and more than 30 in width, was the largest building in the colony: 'They are noble works and quite eclipse our surface show.' Some commentators considered it too well-finished compared with some of the Swansea works and, with its classical façade, it looked quite the part of a 'chapel' of industry.

The grandness of the enterprise was obviously still rankling with locals some months later. The *Register* reported that the produce of the Kanmantoo mine was being dealt with very successfully, thank you, by the Messrs Thomas – also smeltermen from Swansea. They might have been just 'creeping' along, with their single furnace, while others had claimed the advantages of their 'patents right, of their fine buildings, and their almost supernatural skill', but the Kanmantoo had no need of the PCC.

THE YEAR OF EVOLUTIONS: TAKING HOLD OF THE REINS 45

Local developments certainly looked promising but Henry could not help viewing these cautiously. They were 'not quite done with the Swansea boys yet', even though he expected to soon ship actual copper direct to India, suggesting Graham should 'only crow in private' for the time being. In smelting the poorer-quality ores, Henry reckoned the PCC would increase the SAMA's profits by 25 per cent. There was still ore going to Sydney and Van Diemen's Land (Tasmania), despite what Henry had earlier said, and the really top-quality stuff was still going to Swansea. He believed, though, that not all three South Australian smelters could survive, long term.

Henry had to fend off Graham's importuning him to provide gossip on top of the welter of business news. He said he would just give Graham news of 'Miss Betsey Burra', though he never stuck to this stricture. He admonished Graham again for being too free in London with mine information and for falling too easily for speculative projects suggested over there. In particular, a project for a railway from Port Adelaide to the mine seemed very favoured but Henry forcefully cautioned Graham about involvement: 'It won't do now. If you can get the "Britishers" to do it, why very well, but don't you promise anything towards it. Smelting our ore [in South Australia] makes all the difference.' He added that there would not be a third of the ore carted, compared with previously. It was now the PCC's responsibility to deliver pure copper to the SAMA to the value of the ore received from the Association. There would be no economic nor practical point in SAMA shareholders investing in a railway and, as Mel Davies (1977) has convincingly pointed out, there never was. The SAMA did not even have to deliver ore to the smelter, whose own drays had to come and collect a given monthly tonnage.

It was said that the SAMA had placed Walters and the PCC where it wanted them, and it was true that, if the Burra had gone the way of the Princess Royal, the Kooringa Works would have been left high and dry. However, though Henry would try to gain further advantage at later seven-yearly renewals of the agreement, he clearly felt at the outset that it was a *mutually* beneficial arrangement. He was even coming to refer to the PCC as 'our copper works', whereas, in a sense, both the Bunces' and Pennys' operations were more naturally linked to the SAMA through their owners. Indeed, the Bunces were angry with the SAMA directors for having considered an agreement with the PCC at all.

Samuel Stocks Jnr. was connected with the Bunces' project, and worry about it may have hurried his decline, though the erstwhile Resident Director's formidable drink habit had a lot to do with it. Stocks may have been drinking, not just to forget his business worries, but to escape domestic embarrassment. Stocks had suffered four fits on the arrival of some woman to whom Graham had promised a governess' post at a salary of 50 guineas a year. It seemed she had been destined to tutor Stocks' children, but the salary was now beyond his (Stocks') means. John Adams reflected: 'Beware of Women and Edge Tools', a little cautionary aphorism to add to one he had already given Graham on his departure: 'Be careful of women and wine as you would loaded firearms.'

At the mine, Henry made further improvements. Despite all the other anxieties during October 1848, he had succeeded in persuading Adelaide chemist William Elphick to take the post of resident Assayer, so they would know for certain the value of their ores, as compared with Burr's fumblings. Elphick soon earned a pay rise.

Captain Roach's reliability was also rewarded with a pay increase, to £26 a month. This, perhaps coupled with an ever increasing burden of responsibility, prompted a bold – perhaps foolhardy – attempt by Henry to have his own salary raised quite sharply. The burden of work upon the young Secretary had steadily mounted over the previous couple of years and, with his family growing, he took an enormous risk which might well have ended disastrously. He read his request to a regular board meeting, both for a rise and a clarification from the directors of how they saw his future. The directors considered a rise from his £350 to £500 reasonable, effective from the following month. Henry had not yet finished, however, and promised them his answer at the next board meeting. At the close of its business, he had a quiet word with Chairman Charles Beck. His demand was now on a much grander scale: he wanted the board's offer back-dated for the last couple of years, and then £600 going forward. Smelling a rat, Beck asked Ayers if he had had an offer from G.S. Walters, by any chance. Having been lampooned in the *Mercury* for falling prey to Walters' business cunning, Beck and his fellows must have felt caught off guard. Being the soul of discretion, the Secretary answered that he could not possibly comment as it might compromise his position. The directors decided that the chairman should formally ask about any Walters offer but, again, Henry insisted on remaining tight-lipped. The directors resolved that it would be discussed at the next meeting with, meanwhile, all forbidden to talk to their Secretary on the matter. The nail-biting had another week to run; the 28-year-old Henry was really chancing his arm, perhaps having seen an expert negotiator like Walters at work – possibly even egged on by Walters, who had been writing to him privately, in any case, about the lack of an agreement between their two concerns. Despite having come to lean more and more heavily upon their Secretary, both for day-to-day handling of the SAMA's affairs and therefore the accumulation of their wealth, the directors chanced calling Henry's bluff, teaching him a lesson about indispensability. Their counter was a point-blank refusal of any increase and, into the bargain, they deprecated the whole tone that he had adopted in trying to get one.

Now, we cannot know whether Walters had led him to believe there might have been an opening for him with the PCC but, on balance, it looks more likely that Ayers was just trying it on. Adams attests to the fact that Ayers knew, or believed he knew, what Walters himself was on and, if true, Ayers was looking to the SAMA Board for half that – with his responsibilities, he might have thought, a very reasonable figure. Despite the rebuff, he had to swallow his pride and soldier on, for not a penny extra in the short term. Only after he had been given time to stew, while his office subordinates were granted rises, did the directors let him have an increase, and then, only to the £500, not the £600 he had desired. This put him on a par with Thomas Williams, Superintendent of the PCC's Kooringa smelter, but nowhere near the level of former Superintendent Burr who, Henry observed, was down on his luck, 'poor devil', unable to get a post anywhere.

Henry spent an entire winter fortnight at Burra in July 1849 looking into everything, remarking 'By St. Paul, the work goes bravely on!' He had acquired a few more shares himself: 'I shall never let these seductive shares alone. Nothing else tempts me, Rising Townships! Rural Sections! Marine Villas! Railway to the Port! None have charms to compare with my old friend "Betsey."'

This personification of the mine as 'Miss Betsey Burra' seems to have stemmed from a letter to the *Register* purporting to be the reply of the precocious young lady to some criticism of 'her' by a Reverend gentleman. Perhaps Ayers was the author: his sister was known as Betsey. Shares in 'Miss Betsey' might indeed have held the greatest appeal for Henry, but he had not held entirely aloof from the other areas of investment he decried to Graham; he had already put down for shares in a British company proposing to build a railway from the city to the Port. Henry was also listed as buying sections of land, though he may have been acting on behalf of Kingston and Paxton, and not for himself.

Henry further expanded his SAMA duties, having been 'invested with the order of the hammer' – becoming a licensed auctioneer – in order to hold periodic sales of copper at the city office.

Coke at the Coalface

At the SAMA's city office, Henry's hand was firmly in control, and he was steadily forming a more than competent team on site at the mine. Captain Roach was well seconded on the administrative side by William Challoner, who was pretty smartly promoted to Accountant. With Elphick as trusty Assayer, the team which would see the mine through its most expansive years was nearly complete.

In view of all the local smelting capability, Henry let the London agents and James Coke's associates, Leach, Richardson in Swansea, know that their ore-shipping services were likely to be required less. This may have been the catalyst for Coke to capitulate to Henry's blandishments and become a SAMA employee, meeting the need for a reliable mine storekeeper. It may be an indication of the closeness between the two men that Henry's letter engaging James was copied into the Correspondence Book, used for copies of letters to existing mine officers, whereas all similar letters were copied into the giant General Letter Book containing correspondence with non-mine individuals and companies.

James took up his post, in the SAMA's former smelting house, in late October 1849. There was possibly a split between James and his former boss Younghusband, though relations may not have broken down completely, because about 18 months later, Anne Ayers and Mrs Coke were 'at Mrs. Younghusband's', seemingly the best of friends. Around the time of the Cokes' removal to Burra, Mrs Younghusband must have been facing the torture of the damned because of a scandal in which her husband was involved which made the papers.

Ex man of the sea, allegedly with a nautical roving eye to match, William Younghusband Jnr. had failed to protect the good name of his friend Osmond Gilles' daughter. The alleged philanderer freely admitted that he had spent hours in her bed-chamber, and they had told Mrs Younghusband this, shortly afterwards. Miss Gilles alleged that he had clasped her hands and addressed 'lewd and unchaste expressions' to her. It was suggested that she had not repulsed him as firmly as a person of her maturity might. His story was that it was she who had fetched him to her bed-chamber where he had tried to comfort her in her hysterics over the threat of her then fiancé, Dr Moreton, to shoot himself. The new Mrs Moreton could not swear that she had not fetched Mr Younghusband, and the newspaper editor observed that hers was not the ideal behaviour for young maidens to emulate – they should neither 'dally with temptation nor parley with the Devil'. It was clearly a very messy case, Younghusband's extenuating claim being that a maid had been present the whole time, and with a candle lit, but a jury found against him. No doubt Mrs Younghusband shed many a tear on Anne Ayers' shoulder all through this period, and on Mrs Coke's, too, before she left for Burra a week later. One wonders how many eyebrows were later raised when it was William Younghusband, as Chief Secretary in the South Australian Government, who introduced a parliamentary bill to enable divorce.

It may not be wondered at if the rather pedantic Coke accepted the SAMA appointment at least partly to remove himself from the bad smell around

Younghusband's. The Adams had been similarly aghast at the two-timing behaviour of the wife of J.B. Graham's erstwhile partner, Michael Featherstone, with their driver. Until they cut Mrs Featherstone, the Adams bore society's cold shoulder.

James had already done little services for the SAMA directors like ordering coffee mills for them through his Swansea principals. What Henry probably hoped to get from his appointment as Storekeeper and Bookkeeper was the application of a tidy, indexing, accountant's mind to the management of the stores. Henry was less inclined to hold the reins tightly from the city office, with regard to James Coke, than he was with the generality of the mine's officials. James' longest-serving predecessor in the post, S.W. Humble, had had to wear criticism from Ayers for failing to check items into stores with much rigour. Between these two, a short-lived holder of the office virtually lived up to his name – Edward Strike. Relegated to a subordinate role under James, and despite encouraging prods from Henry, Strike still failed to pull himself together sufficiently to get his books up to date and went out the door.

Coke, the energetic new broom, asked Henry to order new ledgers and records books, laid out in more clearly defined columns and categories. He also soon complained about Bell Freeman, the SAMA's preferred carrier for major items such as pieces of the massive beam-engine needed to pump out the deepening mine-workings, as well as precious items such as the samples of the almost gem-quality malachite ore prepared for London's 1851 Great Exhibition. Freeman's status as a veritable 'peer of the realm' of draymen cut little ice with Coke. The new disciplinarian complained at Freeman's unloading items willy-nilly and thus 'deranging the Stores'. Doubtless, Henry received James' report with considerable satisfaction. It would have appealed to his own organizing mind.

Family Ties

Henry had put his faith in James regarding much more than the efficient running of the business, though. The Ayers' eldest, Frank, aged about seven, was living at Burra with him. It may have been because Anne was expecting her next child that Frank was 'farmed out'. It is unlikely that she was short of domestic help. They could afford service from 1846 and it was a period of brisk arrivals of immigrant vessels but, with four children under seven, home life would still have been demanding. Henry was prompted to mention the subject to Graham because he had gathered that Louisa was expecting their first child. As Henry put it, he was 'on the lookout for [his] no. 5' and would soon need a bigger house. He went on to add that little Frank had been with the Cokes for so long in Adelaide that they could not bear to separate when the Cokes moved to Burra: 'the Boy is extremely happy with them and highly interested in all the mining operations'. Does one sniff an aspiration on Henry's part to father a dynasty of Ayers at the helm of the SAMA?

The closeness of the Ayers and the Cokes is further underlined when the Ayers' 'no. 5' was named Charles Coke Ayers. Sadly, he was the first of Henry and Anne's children who failed to survive long. The unhappy father noted the date of death in his *Memorabilia* as 3 October 1850. He wrote 'Aged . . .' but could not fill the blank space. Little Charles had in fact survived almost five months. Henry's immediate task was to secure a burial plot, and a form survives registering the details of a 99-year lease of a West Terrace cemetery plot, dated the day of the loss and bearing Bishop Short's signature. If Frank's return from Burra had not already happened, perhaps Anne would have wanted her whole brood around her at that unhappy time. As James' brother-in-law, Captain Powell, put it when talking of his large family, 'we are a family of eight; ten if you count the little angels that are gone'. The Ayers were perhaps lucky, and probably unusual, in that their first four children had survived and thrived thus far.

A Greek Tragi-Comedy

Controversy over ore values re-emerged as an issue for Henry and the SAMA by the time James took up his post. The PCC's 'analytical chemist', Glaswegian Andrew Thomas, whether by his own design, or under instructions from management, consistently assayed the ores at slightly lower percentages of copper content than did the SAMA's Elphick. This would mean if unchallenged, of course, that when it came to the PCC's returning pure copper to the SAMA as payment for ore received, the SAMA would be entitled to less copper than it had bargained on and the PCC would have more to sell on its own account. This was a game, however, and a novice player like Henry had yet to master the rules.

These persistent discrepancies became a source of irritation to Henry, who went to the extent of ordering new weighing-scales, from Avery of Birmingham, so that William Elphick would have the 'identical scales to Mr. Thomas', whose assays were 'not always dependable', in Henry's euphemistic phrase. With a deal between the SAMA and the PCC still unsigned, aggravation was such by September 1849 that Henry was writing to Gregory Walters to say that he was willing to meet him in a Court of Equity, if that is what it would take to get agreement finalized. At the same moment, though, Henry was also giving seemingly friendly advice on how Walters should go about purchasing land to develop a little township at his port at the head of

Gulf St Vincent. When Governor Young saw the sad state of the mangrove swamp that was to be named Port Henry, he was reluctant to lend his name to it. Walters' development instead commemorated the originator of the idea of the colony of South Australia, Edward Gibbon Wakefield, and Welsh coal came in and copper and some good quality ore went out, to the Spitty works, through Port Wakefield.

Walters had a further perfect excuse for not ratifying the agreement with Henry, due to ore 'washing'. As the chimney stack of 'Roach's' pumping engine belched smoke on the western horizon, Walters and Williams could not miss it from the smelting works, nor its significance for their business. The water it drained from the mine workings, edging below the water-table, itself became a tool for the better 'dressing' of the ores at grass. The PCC was limited to the Burra mine's lower-quality ores, while the Pennys were allowed better ones, at over 35 per cent copper. Now that the mine had the water to 'wash' its ores, to enhance the copper content, Walters complained that the PCC was being disadvantaged.

The giant headache inspired Henry to let off steam by devising a dramatis personae suitable for a Greek tragedy. It could have come from his pen any time during 1849. His chosen title was *Mythology of the P.C.C.* The setting was Tartarus which, as the lowest part of Hades, or even the place below Hades, was a suitable synonym for South Australia. Pluto and Proserpine he cast as Sir H and Lady Y[oung] and, as Pluto was at least partly the bestower of gifts hidden in the earth, it was probably apt, especially as Governor Young had bestowed that mineral wealth tax-free. Tantalus seemed to fit Mr Schneider; his source of refreshment perpetually out of reach was 'The Cup formed of an expected Indenture Witnessing'. Sisyphus, '(a good man)' was 'The Manager of a Certain Company rolling up the B[urra].B[urra]. Hill a Stone of Malachite'. This clearly represented his own thankless and never-ending task managing the Burra. Prometheus, yet another 'good man', was 'The Superintendent [Captain Roach] of d° [ditto]. his sweet bread tormented by 9 vultures & 1 Secretary Bird'. For the Furies, Henry cast the lawyers for his own side: Gwynne, his partner Lawrence, and James Hurtle Fisher. The 'Good Fury' is R.D. Hanson, the PCC's lawyer, whom he describes as a 'friend of Sisyphus' – possibly ironically, though they may really have been friends by then. Cerberus, Charon, Shades Below, and others figure in the cast list, and the River Styx is, of course, the Torrens, which is 'full of Sticks'. Minos, inferior deities, Burning Heat, Imps, Crawling Things, Stagnant Pools, darknesses, etc., etc., 'Eadem, eadem, eadem, in Regionibus ambobus [likewise, likewise, likewise, in these Regions together] including bullock men, Smelters, Laborers, etc.'. Birds tormenting his liver had to remain anonymous, but numbered nine – the directors – plus one 'without wings'. Perhaps the latter was the non-avian 'boar', Burr, or even the absent, yet influential, Graham. Intriguingly, the Elysian Fields are to be found in Light Square – the less-than-respectable home of what Adelaide nightlife there was!

While Henry wrangled with the PCC on the grand scale, James dealt with the day-to-day trials. He brought his accountant's finicky regard for detail to bear upon the logging of the ore qualities being collected by the smelters. He requested a ledger which would allow for noting tenths of a per cent; even a stickler such as Ayers felt this overly particular, suggesting that an accuracy to the nearest quarter of a per cent as quite sufficient. The two men's relationship shows itself clearly as something unusual over this because Henry is prepared to be cajoling and accommodating, whereas for any other officer, he would simply have laid down the law. In the meantime, Henry had a quiet word with the Second Assayer at the mine, suggesting he adhere to the customary quarter per cent for the moment. Not many years passed before Storekeeper Coke's position was partly vindicated; noting units to an eighth of a per cent became the norm.

Even if relations were less than easy between the SAMA and the PCC, Henry instructed his storekeeper that he should help the smelter with any of the mine's stores that could be spared, if there came an urgent call from across the creek. The PCC had agreed to reciprocate. Only Roach and Coke were allowed to oversee these lendings and borrowings and James was strictly enjoined to ensure that they were recorded.

Bachelor Roach spent the 1849 Christmas season at the Ayers' and they all enjoyed a pleasant evening at Prospect House with the Adams. Speaking of the Captain, Henry described him to J.B. Graham as the 'same energetic fellow he was and beginning to show gray hairs in the service'. In high seasonal spirits, Henry had been teasing Roach that he should get married, while, on a more sombre note, he told Graham 'Poor Sammy [Stocks Jun.] is no more.' He had become 'prematurely old and could scarcely walk or ride in a Gig without being supported. What a sacrifice of human life!' Henry and many associates published a protest in *The Mercury and South Australian Sporting Chronicle* at the *Register*'s obituary notice which revealed more of Stocks' 'private character' than they thought the man had deserved.

Continued Good Prospects

New valuable lodes were being discovered, old ones were still producing well, the PCC was getting along famously, and copper was being sold as fast as it was produced. Yet, an ominous sign appeared by the February of 1850.

Henry wrote Graham of the affliction of numbers of the Burra workforce with 'Californian Fever' – a shipful had just sailed for the gold-fields, where a rush had begun following the discovery of the precious metal two years earlier. On the brighter side, more new men were arriving, though 'very few of the right sort' in Henry's estimation – whether they lacked the right mining capabilities or were simply not prepared to work for next to nothing is not clear. In any case, Henry wondered if Graham could encourage more miners to emigrate, and he may have been envious of the success the PCC was having in increasing its workforce; a 'bevy' of new men, as Leach, Richardson put it, had come out on the *Glen Huntley* with the *Union* not far behind. John Stephens editorialized that the 1848 Revolutions had been at the root of all the difficulties of late 1848, but the grinding of the men had stored up problems for the SAMA: experienced men had left for California and new arrivals from Cornwall found they were arriving to lower wages than they had been led to expect.

With J.B. Graham resettled in Britain, Henry sought ways of sending his dividend payments to his friend's best advantage. He also had the uncomfortable duty of informing Graham that the directors wanted him to step down as a SAMA Trustee, a position to which he had been relegated on giving up his directorship. The distance was causing inconvenience. Henry was careful to reassure Graham, though, that this was no kind of slur upon him. Ayers would try to keep Graham happy, and well-advised, over the years, to avoid any rash behaviour by the far-away major shareholder that might upset the entire Burra enterprise.

The warmth of the relationship between the SAMA's Secretary and the now London-based ex-director is underlined by the presents sent out by the Grahams to the Ayers. In April 1850, Henry wrote to say how pleased his boys were with a box of toys, dancing and capering round it while it was being opened with a gusto that 'quite put a corrobiree in the shade'. Henry was amazed at the improved quality of such articles – particularly a doll – since he left England. The stiff and straight Ayers, of the formal SAMA correspondence, revealed his inner child when he confessed to spending two evenings occupied with the building bricks. He also wished to thank Louisa for a set of mantelpiece ornaments and a copy, in her hand, of Graham's journal; Henry had been permitted to read it for the Adams and jokingly warned Graham: 'I know all your secrets and all about you so beware!'

It was obviously a great source of concern to Henry that the pool of possible SAMA directors was diminishing. Some existing directors were being drawn back to Britain on the strength of their dividends, including John Waterhouse and Charles Beck. Henry freely admitted again that he should like a holiday, himself, but he could see no opportunity. It was becoming necessary to reduce the number of shares held in order to qualify as a director, from 20 to 11, as the number of South Australian residents with a large shareholding was diminishing. In Henry's opinion, this was less objectionable than reducing the number of directors who, he felt, were already being drawn from a poor choice of candidates.

At the purchase of some new mineral land, the Snobs had been 'successful again' in getting what they wanted. The question had been 'how to get the Burra into it who are the people most of all feared?' – an indication of the increasing power of the SAMA as perceived by other local business interests. They would need more men in order to work this new acquisition and, again, Henry prompted Graham to see what he could do to promote the emigration of miners. Henry admitted that the SAMA's success had 'left a lot of "soreness" on behalf of the unsuccessful with the Burra interest'. Thomas Waterhouse would replace his departing brother, John, as a director, and William ('Mr. Billy') Sanders would replace 'poor Sam Stocks'.

In early June 1850, Henry noted from the newspaper that the Grahams had a 'son and heir'. Little Henry might have been named in honour of Ayers, or of Louisa's brother, Henry Rymill, who would come to figure prominently in Ayers' story. Henry reported continued good prospects from the mine, which made Roach less hesitant about taking on a wife; Henry was optimistic that he [Henry] would soon be arranging a wedding. On a more serious note, he wanted Graham to harry one of their London agents over a second, more powerful, pumping engine, already on order. As he put it, his fingers 'itch[ed]' to go from the '50' down to the '60' [fathoms]. The 'bread and butter of many thousands depends on the prosperity of the Burra Burra'.

At the same time, ever cautious Henry put it to Graham that, when possible, he wanted the directors to set up a reserve fund to 'guard against a rainy day'. He went on to discuss at length some plan of a London agent, under which shares would tend to gather back in Britain. This was not calculated to find favour with a committed South Australian, which Henry, through force of circumstances, was fast becoming.

5

ALL THAT GLISTERS . . .

From early 1850 until mid-1853, crucial years in the life of the SAMA, the mine and Henry Ayers, there is a frustrating lack of personal letters. The official SAMA correspondence indicates how life was turned upside down for our protagonists and, from his own *Memorabilia*, we can trace what appears self-satisfaction on Henry's part, when he took stock of his situation. In his methodical way, on the last day of 1851, he noted down earnings and expenditure since arriving in the colony, coming to the pleasing conclusion that he and his family were then worth '£5,795:0:9' – 'By steady industry and the careful savings of Eleven years'.

Scrupulous record keeping had become a necessity, if not an obsession and, from a recently unearthed box of original papers, we find Henry had not only lent money to former boss J.H. Richman, but also to the likes of (police) Inspector Tolmer and – to the tune of almost £3,000 – SAMA Director William Sanders. In addition to his quite generous salary, which by September 1850 had reached his desired £600 per annum, he was saving with the SAMA, at a return of 6 per cent.

It is somewhat glibly asserted in brief biographical paragraphs on Ayers that he became the SAMA's Managing Director in 1850. Nothing regarding such a transformation exists in the minutes of board meetings or that year's AGM. As ever, 'Secretary' was how a *Sydney Morning Herald* journalist described Henry in 1851, going on to repeat as mere tittle-tattle that he was believed to have possibly too much influence 'among the Directors', as though he were one of their number. He appears to have bought his first six SAMA shares at some stage before the end of 1845, though Ian Auhl (1986) asserts he held 11 shares from the SAMA's inauguration. He would build up or reduce his shareholdings over the years, with the vagaries of the market, but was never a director as such. A managerial role was nevertheless emerging. Captain Roach had undoubted expertise in the practical aspect of running the mine, but it needed someone to co-ordinate the entire business strategy – dealing with Mr Walters and the PCC, and overseeing the sale of copper in Adelaide, Britain, India and China.

Henry had taken few decisions on his own responsibility in the earliest days, apart from engaging labourers, or little things like the impromptu purchase of blasting powder, which the directors would afterwards endorse. His responsibilities steadily grew, as with the purchasing of land, though he was given budgetary limitations. More significant was his prominent position in negotiations with Walters over arrangements with the PCC. After a couple of months, he simply reported on the negotiations to the directors.

One reason for the weight of decision making falling increasingly upon him was the fact of weekly board meetings; there was an ever-growing correspondence to be dealt with, with the mine's increasing complexity in terms of personnel, and dealings with overseas as well as local agents. Decisions and letter preparation were required with minimal delay, though Henry invariably read letters over to the directors before sending them. His grasp of the fundamentals of the copper trade, gained quickly on the job, was first-class. There were generally no quibbles by the directors over his decisions, though that may not have been a great endorsement, since most had no previous experience of the copper trade. Yet, Henry was certainly making a great success of the enterprise.

A second reason for his centrality to the ongoing success was the fact that the board was regularly changing, into the 1850s, and directors' attendance at board meetings was not outstanding either. With mountainous dividends rolling in, various board members were in the happy position of being able to afford the time and cost of a visit Home. Henry was a still point around which the moving forces of the Burra constellation revolved. Thus, he regretted to Graham that he could not foresee any immediate or mid-term possibility for his own trip Home.

Drought and Flood but Little Sympathy

The *Sydney Morning Herald* regaled its readership with a detailed appraisal of the Burra enterprise in a three-day series of articles. This celebrated virtually the peak of the mine's success, had the writer but known it. Severe summer drought, denying the draught animals either water or fodder, had put serious pressure on the PCC's ability to keep its furnaces fuelled, as coal could not be brought up from Port Wakefield, or wood from the Murray plains. Not for nothing was Gregory Walters nicknamed 'Walters the Woodless'!

The southern autumn added insult to injury. It started encouragingly in early May, the *Register* speaking of 'the late warm rains'. The paper soon changed its tune to a 'destructive flood' and Henry heard of people washed out of the Creek; he wrote to Captain Roach, instructing him to 'afford them all the accommodation in your power'.

Henry's initial sympathetic response gave way to a hard-headed realism. Once the community had had a couple of weeks to recover, he instructed the mine Accountant, Challoner, to place a notice on the smithy door. Any Creek dwellers still preferring that mode of accommodation, come 1 December, would cease to be SAMA employees, and treated as trespassers.

Challoner barely had time to get his notice up before a repeat performance ensued. On 11 June, the *Register* reported 'another great flood at the Burra'. Three days before, rain, hail and thunder 'like another Niagara' had begun at six in the evening and was still going at six the following morning. Losses of cash in the 'Savings Bank' – another nickname for 'Creek Street' – ran from £50 in some cases to as much as ten times that.

A determined return to cave dwelling after the first flood is evident from Henry's injunction to Challoner, and it is fortunate that loss of life was restricted to just one person. The death of widower William Box encompassed a double-edged moral tale. On the one hand, he had been helping his neighbours to secure their furniture. The Good Samaritan was killed when the roof of his dug-out fell in on him while he then tried to rescue his own belongings. He left three young children, and it was lucky for their sakes that the police managed to recover an alleged £400 of Box's savings. That was the other side of the moral coin; he and his young family had been living in the primitive conditions of Creek Street while he actually owned the leases of several Burra cottages.

Henry was in no mood for sympathy a second time. No help was to be given to those whose homes and belongings were again washed away; the mine's requirements came first. No SAMA's stores were to be used to assist the homeless. Henry had no patience with men prepared to risk their families for the sake of a few shillings a week in rent money. If he thought about it more coolly, he might have considered there was something he could have admired in poor Box, the entrepreneur. At some sacrifice to himself and his motherless children, Box had been trying to improve his situation by renting out cottages. It was just that Henry and the directors were unhappy with anyone who was avoiding paying *them* rent.

Henry knew his Captain Roach sufficiently well to realize that his hard-hearted injunctions towards the washed-out families might be softened. He admonished him to be firm with the people. His duty was to the mine, whatever his feelings, and Henry took full responsibility personally, though he was sure the directors would concur with him. Here he was, at 30, brimming with the confidence born of his own success, surveying from afar, and from on high, the folly as he saw it of the independent-minded miners.

Having told Roach to withhold further assistance to washed-out families, Henry conversely instructed him to assist the PCC in any way practical to rebuild its damaged bridge crossing the creek between mine and smelter. Henry's sympathy and patience in that quarter were also short-lived because, a month later, he insisted the PCC fend for itself. He was irascible, too, with the populace's complaints that a town bridge for the Burra Creek had not been completed. Another of his letters landed on Roach's desk saying it was no wonder the bridge was unfinished as the citizens had not come up with their promised quarter of its cost. Thinking perhaps better of it, he told Roach work could go on with it provided it did not hold back progress on the new engine-house.

Perhaps the only benefit from the destruction of the dugouts, remarked the *Register*'s editor, was that the calamity would not happen again. The combined impact of the flood, and Henry's diktat regarding cave dwelling, may also have produced the benefit of reducing illness within the community, which the insanitary living conditions of the creek-caves had promoted.

There was housing available beyond the bounds of the SAMA's Special Survey area, but the directors, through Ayers, had acted early to try to make sure that this did not give the miners a more reasonable option. When the government put Crown land up for auction just beyond the SAMA's boundary, a year after the miners' strike, this new village fell largely into the SAMA's clutches. In 'inclement weather', quite a crowd turned up outside the police station in Kooringa to hear Henry fight off other bidders. Ridiculously, the 20-odd acres of 'Redruth' raked in for the government well over what the entire 1,040 town-acres of Adelaide had fetched 12 years earlier. Henry, for the SAMA, mopped up nearly 80 of the lots.

The 'Voluntary' Principle

Whatever problems drought and flood had brought the smelter, the Burra mine had been doing well in early 1851. Californian gold had proven an insignificant threat to workforce numbers and there was even an increase in take-home pay through overtime. Presentable cottages had been built by the

SAMA, to Kingston's designs, in Thames Street, Kooringa. He also provided designs for a more basic row of cottages, built by Robert Henderson, forming the western portion of a square named after SAMA Director William Paxton. From one historian's perspective, Ian Auhl (1986) saw this as an example of the directors clawing back what they could from their workforce, along with medical subscriptions for a doctor and small hospital; Geoffrey Blainey (1993), by contrast, seemed to question whether the directors should have been running a mini welfare state a century early.

Various denominations applied to Henry for leases of land on which to erect their chapels and churches, one of which served the 'Bible Christian' sect of the Methodist Church, which built a solid little chapel, backed by its manse, on the southern end of Henderson's Row. This charming little edifice, as stoutly buttressed as if it were a Gothic cathedral, has transformed into the reception centre for today's visitors who take holiday accommodation in the spartan, yet cosy, Paxton Square Cottages. The Reverend Pollitt – the best Anglican preacher in the colony, according to Mr Adams – was refused a grant of land for a school which he wanted to build close to the Storekeeper's cottage. Henry let it be known that the SAMA had no intentions whatsoever of 'giving away' any part of its property, even in the best of causes. He also reserved for the SAMA the right to any mineral finds on leased-out land.

The question of whether the state should provide financial aid for the building and support of churches, especially in thinly populated country districts, was hugely controversial. The even bigger question of whether the Anglican Church should be an established church in South Australia, as in Britain, generated heated debate. It was bad enough for many that a Colonial Chaplain – of course, Anglican – had been appointed from the start of the colony. A great number of Nonconformists had seen the colony as a refuge from the influences of the established church at Home. Pious Anglican Governor Robe had run into a storm of controversy when he put forward a law for supporting church building, and the maintenance of its clergy, in 1847.

It is not clear where Henry Ayers stood on this matter; we do know, though, that as close an associate as George 'Paddy' Kingston was vehemently anti-state aid for religion. A strong Anglican, though Irish, he nevertheless did not wish to see his particular creed underpinned by government funds. Given Henry's latish conversion to more active churchgoer, it is highly unlikely that he would have supported state aid. The 'Voluntary' principle, where churches, chapels and clergy were supported through the sole efforts of their congregations, seems more likely to chime with what we can tell of Henry's outlook in this period. Indeed, he and Director Captain William Allen would patronize the building of the Anglican St Paul's, Pulteney Street, in 1856.

Candidates Come Forward

The Voluntary principle again came to the fore in mid-1851, becoming perhaps the main concern of voters in the first elections to the new South Australian representative assembly. The Westminster parliament had seen fit to grant a form of representative government to all the Australian colonies, except Western Australia, in one go, believing that satisfying colonists' aspirations for a degree of self-government would translate into a greater sense of loyalty within the Empire. South Australians were dubious about three aspects of this paternalism, however. It resurrected the question of supporting an established church. Their precious land values were also under threat; the colonists feared that a market value, as in New South Wales, would replace the guaranteed 20 shillings an acre that the Systematic Colonizers had considered vital to the success of a more genteel South Australia. Finally, what seemed on the face of it a sensible proposal of the Mother of Parliaments, to have a periodic General Assembly of all the Australian Colonies, was perceived as a probable infringement of the sacred liberties so prized by the founders of South Australia. This prim colony, with a large Nonconformist element, free of the taint of convictism, jealously guarded its independence of the other Australian colonies. The candidates for election, then, had two principal questions to answer: were they for or against state aid for religion, and for or against the immediate radicalization of the newly granted Constitution? This was famously lost for days in the captain's laundry locker of the ship bearing it to the colony.

Again, we have no idea where Henry Ayers stood on these matters, nor what interest he took in the political jamboree running up to the election, coinciding with the second Burra flood; he expected 'broken heads' as a matter of course – in the grand British electioneering tradition before votes by secret ballot. All one can say is that numbers of his close business and personal associates were candidates and indeed became members of the one-third nominated, two-thirds elected Legislative Council. The likes of Kingston, Peacock and Younghusband – the scandal of two years before seemingly forgotten, or forgiven – wound up in the single-chamber legislature, not to mention the SAMA's lawyer, E.C. Gwynne, and the PCC's, R.D. Hanson. Henry's letter to Captain Roach concerning the election was ostensibly clear-cut in stating that the SAMA directors had no interest in how their workers voted. Whether Roach was expected to read between the lines is open to conjecture. The newspaper reports certainly suggest that Roach indicated his preference, at the lively public meetings in Kooringa, by leading the applause for Paddy Kingston's speeches. Kingston fought off a challenge from his opponent, Henry Mildred. Kingston's involvement, as in so many other spheres, provoked controversy and Henry wrote to the director, on behalf of the SAMA, feeling that Kingston owed it to the Association to take whatever steps necessary to 'clear himself from the imputation cast upon his character by Mr. Mildred at Kooringa, by stating that he [Mildred] convicted Mr. Kingston of three deliberate falsehoods'.

One alleged falsehood was that Mildred was in gaol and that was the last that the Burra voters would be seeing of him, but Kingston denied making this allegation. Kingston also faced questions about the SAMA's seeming determination to buy up the whole of Redruth at the auction, two years before. Kingston stood up for Henry, insisting that Ayers had only bid against the city-interests, themselves determined to acquire the township; he had not bid against any bona fide working man. At least, that was Kingston's story.

The Deputy Grand Master of South Australian Freemasons maintained his dignity against the challenge of the Worshipful Grand Master, amid veiled accusations of bullying and bribery. William Paxton was believed to have put £100 towards buying Kingston the seat – no doubt most easily reaching his targets through the voters' thirsts in the four Burra pubs Paxton owned.

Whether Captain Roach had much influence upon the local result is questionable, in any case, as the vast majority of the workforce was ineligible to vote. Their three-shillings-a-week rentals of SAMA cottages failed the basic qualification of owning property worth £100, or paying at least £10 annual rent. Even if men were ineligible to vote, they showed a marked enthusiasm for attending meetings and assessing the arguments for and against further reforms. Like other South Australians, some were for and others against 'Chartist' notions like universal male suffrage, vote by secret ballot, and short parliamentary terms, none of which was available to the voters in this 1851 election.

Before the end of July, a dinner at the Burra Hotel celebrated a Kingston victory. It was a six-hour jamboree of eating, drinking, speeches and song, with Henry a keen participant in all four. Local tradesman T.W. Powell, in the chair, praised the election of Kingston as putting in place the keystone that completed the arch on which the success of Burra was founded – the workers on one side and the SAMA on the other. Powell overstretched his imagery a touch in raising a toast to Henry Ayers, as another keystone for that arch. He was a 'trusty friend in deeds as well as words' and Powell raised his glass 'in friendship, love, and truth' to cheers from those present. In addition to responding to the toast, Henry sang them 'Have you not heard of the Monster Mine?' in, says the reporter, 'excellent style'.

Gold Trumps Copper

The team Henry had assembled for the smooth running of the mine lost a valued member by mid-1851, when James Coke moved into the grain and milling trade. Perhaps this partnership was too attractive, or life in a raw mining town too uncongenial. Possibly James' ineligibility to vote, through having free accommodation at the mine, was a catalyst; he had a keen interest in public affairs. Maybe, though, the myriad responsibilities Henry placed on him had just got too much. Months before, brother-in-law Captain Powell had conjectured that he must have been 'dreadfully fagged'. James had had a succession of incompetent clerical assistants, the machinations of his compatriots at the smelting works to deal with, awkward draymen to haggle with, the tasks of a quantity surveyor to adapt to when Henry required it, plus petty aggravations like the refusal of the Pennys' representative to accept an ore delivery on a Saturday. He had even been requested by Henry, if possible, to act as a debt collector for his friend; James was to try to get the £40 out of Gregory Walters, which Henry had lent him one day at the Port. James had no kind word for Walters or Williams, maintaining that George Ewbank, Williams' PCC subordinate, was the 'only gentleman' at the establishment. Henry farewelled James with regret and 'very warm gratitude' for his 'assiduity in the Company's service'. James' departure was only the start.

Gold in California was one thing, but when news of the dazzling mineral's discovery in neighbouring New South Wales reached Burra, the miners were unsettled. Initially confident about retaining the workforce, Henry was soon writing to encourage German migrants. Much greater disruption, to South Australia's very way of life, followed similar news from Victoria. A large section of the adult male population either took ship or trekked overland to seek fortunes on the gold-fields, followed by a bevy of Adelaide's ladies of the night.

The Potts family history group, in their story of Frank Potts and his family's Langhorne Creek winery, narrates an anecdote which, if true, would illustrate Henry Ayers' seizing of the entrepreneurial moment. Potts lore has it that Henry asked brother-in-law Frank's advice and help in putting up a basic inn, the Traveller's Rest, opposite his vineyard, as the cross-country trek to the gold diggings passed Frank Potts' door, en route for the ferry across the Murray at Wellington. If the story were true, it would show Henry as extraordinarily prescient: the liquor licence was taken out in March 1851, five or six months before news of the Victorian gold finds reached Adelaide. Perhaps Henry had already noticed dusty, thirsty travellers from the comfort of Frank's living room passing in the normal way to and from Wellington. It sounds as though a Henry Ayers venture preceding the gold period has been attached in family tradition to the upheavals of the trek. In Henry's accounts for the four-year period to the end of 1850, he had a figure of £40 already earned from his Traveller's Rest investment, and nor was it his only pub. Sounding like a nineteenth-century Monopoly game, his accounts for the three-year period to the end of 1851 show earnings from ownership of the Grand Junction Hotel as well.

In comparison with the eastern colonies' gold, copper was small beer. Coke described the Adelaide of later 1851 as a very different place from the one he had left for the mine, two years before. Neither could he see the smelting works surviving much longer, with even the Schneiders' pockets not deep enough to keep it going. Occasionally, even Henry now expressed pessimism about the PCC's long-term survival, having previously exulted in the SAMA's opportunity to break the Swansea grip on the copper trade. Henry was then taken aback, in April 1852, to hear – of course months after the event – of the reinvigoration of the PCC as the English and Australian Copper Company (E&ACC).

Despite the haemorrhage of manpower to the gold-fields, Henry and the board were characteristically loath to increase wage-rates. That was until Osborne, the Captain (Ore) Dresser resigned, lured away by gold. Rises or longevity bonuses for others followed fairly swiftly, even the office cleaning lady benefiting. With a much-reduced workforce, there was little point in trying to keep the entire mine accessible, so pumping was stopped and the lower levels allowed to drown, sustaining whatever damage would ensue.

Although producing a less glamorous product than gold, Burra could at least offer a more secure living, and a skeleton workforce was content to stay and work the shallower levels. These managed to raise surprisingly good produce. It only needed the E&ACC to continue to make copper, from what was being raised and the plentiful stockpile at grass, and the dividends would hopefully continue to flow. The mine, now, was a far cry from the proud producer whose rich samples had just been put on show at Prince Albert's Great Exhibition in London. Henry had asked for a competent scientific authority to produce written descriptions for the samples, freely admitting that he could boast no technical knowledge beyond a 'commercial insight' into the world of copper. J.B. Graham, with no greater scientific capacity, nevertheless busied himself with arranging the displays.

With activities drastically curtailed, it was not worth keeping the new 'monstrous' engine running. Henry had led the hurling of champagne bottles at the nodding beam, christening 'Schneider's Engine' as it lurched into life. Unlike at the start-up of Roach's engine, the successful engineers were actually allowed to sit down with the directors and officers in the *same* hotel on this occasion. Yet, given the new circumstances, their triumph was short-lived; theirs was but to lovingly watch over the impotent beast, keeping its brass-work bright. No pillar of smoke rose by day to guide the traveller approaching Burra. Henry's earlier push for a fund to tide the Association over a rainy day looked timely and, as the men continued to drift towards the gold, as months dragged into years, he must have had sleepless nights, worrying about what was to become of himself and his family. Perhaps the reckoning up of his savings, on the last day of 1851, had as much to do with reassuring himself that he had enough by him for his family's needs in the looming crisis, as it did pure self-satisfaction. No similar balance sheet marks off any other year in the *Memorabilia*. At 30, he might yet have to skulk back into a junior law office position, that is, if there was still any legal business needing doing; commercial life was at a standstill.

Jamie Chambers tested the Victorian waters by sending his most trusty whip – one-eyed 'George' – down to the diggings. George, who was credited with being able to see more with his one eye than could many with their two, had not much encouragement for any would-be diggers; conditions were 'wretched', with the chances of success rated at one in a hundred, and then only if one could manage 'great watchfulness'. With quotes like 'GOLD IN ABUNDANCE' coming from the Adelaide press, there was not much South Australia could do beyond trying to talk up its own potential for gold finds, but it was all rather futile when set against Buninyong, Ballarat and Mount Alexander.

One-eyed George was holding the reins when the directors' coach, with Henry aboard, suffered a breakdown on its September visit to the mine. The roads were in an appalling state with recent bad weather and it took the old sea-dog, Captain Allen, to contrive a jury-rig to get them on their way. A sapling stood in for the broken shaft, and a spot of nautical lashing provided a temporary fix for a broken axle. One can picture Henry, George and the other directors jumping to, as Allen re-adopted his best quarterdeck manner.

The recent elections had come off reasonably successfully and, with the whole colony now electing at least a proportion of the members of its legislature, some other modes of representation for the people were under consideration. The original city council experiment for Adelaide had come at a bad time. With the population exiting in droves for Victoria, now was quite the wrong time to repeat the experiment and the new plans were shelved for six months. Simultaneously, Henry wriggled out of a push by parishioners of St John's to be a delegate to the Diocesan Assembly. He pleaded that he might sometimes have to be away at Burra, and would not like to take on any responsibility that he could not guarantee to fulfil. Possibly, though, religious activities were not high on his list of priorities.

According to *The Flinders History of South Australia* (Jaensch 1986), the future Lord Salisbury – many years later Britain's Prime Minister – visited an Adelaide 'gradually drained of its population'. At 22, in his black coat and white top hat, the visiting aristocrat considered the colony the most 'Yankee' in Australia, peopled by bankrupt farmers, merchants and shopkeepers, few of whom looked upon South Australia as a permanent home. After all, 'few people would like to cast their lot for life in an oven'. 'Money, money, money' was the burden of the conversation of those left in town, and there was more drunkenness, he believed, than in the convict colonies. Swearing had attained a pretty high standard, too, with one oath he had heard causing him to shudder, even though, as he freely admitted to his sister, he came with a background of Eton and Oxford! The much-vaunted freedom that the colony was proud of was simply the freedom to be 'avaricious and rude'. It was not a very attractive picture of the particular oven in which Henry Ayers had cast his lot.

New Arrivals

Henry may have looked back ruefully on what he had said at Kingston's celebratory dinner. His expectation of a prosperous summer could hardly have been wider of the mark. The Burra enterprise was struggling and, by the November of 1852, Henry and his directors were resorting to a little bribery, in the form of a £2-a-head bounty for each West Country miner the Plymouth emigration agent could send. When the agent's first protégés made it to Burra, by September 1854, Henry complained they were too young – some under 20. Henry said they needed a bit of age on their sides, say 25 to 30 year olds, otherwise they probably had not enough understanding of safe timbering underground. He mused on the fact that you could not give these young 'miners' a trial before sending them over. Did he see the irony that he himself had arrived as a 19 year old, claiming skills he did not have?

Despite the inducements of Victoria, another member of the Lockett/Potts family chanced life in South Australia in mid-1853. Anne Ayers' stepsister, Margaret, arrived from Canada with her husband, Arthur Baker, and their family. They would have found themselves in a very uncertain situation, much as the Ayers had done in late 1840.

Acting in Graham's Interests

By the September, with J.B. Graham confirmed in his intention to remain in Britain, his mother and stepfather, the Adams, decided to return Home. Personal correspondence now survives again, Henry thanking Graham for thinking of the former's 'little ones . . . so often'. The latest presents had not quite arrived as yet, by the *Lady Bruce*, but they were the perpetual topic of conversation in the Ayers household, where the ship had come to be dubbed the *Lazy Bruce*. Businesswise, Henry headed off a scheme for a SAMA London office under Graham's, Waterhouse's and Beck's thumbs. No longer attracted by the possibility of running such an office himself, Henry presumably saw potential limitations on his own room for manoeuvre, as well as the problems that would result from removing the locus of control from the necessary centre of activities.

Henry, whom Graham had obviously given something of a watching brief over his South Australian purse-strings on departing from the colony, was now happy to accept the magnate's invitation to act as his official agent for all his interests there. Henry applied his business mind, immediately, requesting full instructions regarding Graham's South Australian properties. He recommended that those that would sell well be put on the market at once, including Prospect House – 'Graham's Castle'. Henry listed its shortcomings, but the advertisement he ran in the *Register*, into the New Year of 1854, told a more glowing tale. For sale or for rent on a long lease, it was definitely a 'des.res.'.

Nevertheless, no one came near the asking price, set as high as £6,700 – when one considers the £40 cost of Kingston's miners cottages at Burra, or Henry's own cottage in Pulteney Street at £115, one gets an idea of the scale of the Prospect House property, although not large going by number of rooms. Henry sold it to a brewer for £4,300 who immediately onsold it to an 'aspirational' young lawyer – an unhappy investment – helped by a Graham mortgage. So, Prospect House was not entirely off their hands. At an auction of all 'the best furniture', bidding was disappointing, with even the carved rosewood pipe-organ having to be bought in. Ayers could at least reassure Graham that he had managed to sell the instrument to Christchurch, North Adelaide, for £250 over four years.

The entire sale had netted only £1,600. Henry further warned that it was unrealistic to look for further Burra dividends for the time being; the upper levels of the mine were becoming exhausted and it was not economic to work the lower levels because wage-rates were high, even though the world price of copper was also at the upper end.

At this point, one is reminded of the range of Pooh-Bah's advice, depending on which office he was being asked to advise under, in Gilbert and Sullivan's *The Mikado*. Ayers swapped hats, from SAMA Secretary to private agent for his friend. In the latter role, his advice was for Graham to reduce his exposure to the vagaries of the Burra. He should try selling some of his holdings on the London market, or 'quietly' in South Australia. This would have to be done very judiciously, Henry cautioned, as there was a danger of generating a mass sell-off if the mine's biggest shareholder were seen to be reducing his stake. Henry was indeed risking his own livelihood, if things went wrong. He was conscientiously advising his friend of his best interests, despite knowing Graham's record in supporting a string of well-meaning schemes in London which Henry seems to have had little faith in at the best of times. If Graham misjudged the timing of any sales of his shares, the whole Burra enterprise could have been undermined.

The close of 1854 was busy, Henry lamenting his inability to get a holiday in England. The Grahams had moved to Germany, for John Benjamin's health, and Henry again had to find the best way to remit his dividends. The matter of payments caused some friction, with Graham feeling that Henry was not forwarding them swiftly enough. At Burra, with things now looking up, the decision was taken to restart the powerful 'Schneider's Engine' and there was a plan afoot, stemming from the business brain of Gregory Walters, for the SAMA to acquire the Kooringa smelting works from the E&ACC. Henry warned his friend to steer clear of involvement; in his opinion, it did not matter who the owners of the plant were, they would never do well and he, personally, would have no hand in it – it would be 'pure madness'. He expected that the works would encounter carting problems, through drought, once again and the SAMA would simply have to revert to sending its ores to Swansea, if and when the E&ACC went bust.

While gold ruled, the E&ACC had grappled with transport problems, both of drought and the loss of almost every drayman to Victoria's diggings, by resorting to panniered mules. Gentleman George Ewbank got this idea from seeing the South American copper mines, and the E&ACC shipped them over, with their Spanish-speaking handlers. Home-made mule wagons proved an even greater success, despite the reservations of doubting Thomas Williams, who had nothing but complaints against animals and handlers. Perhaps, if Henry had been more of a risk-taker in the business world, he, Graham, and others might have increased their profits by becoming part-owners of the Kooringa works. No doubt, risks were all he could see at this period, for he would have been well aware of the massive problems of transport, labour shortage and, particularly, the dearth of skilled workers the E&ACC now suffered.

The wheeze devised by the acting manager of the Bank of South Australia, George Tinline, of sending a police escort, under Inspector Tolmer, to the Victorian gold-fields to accompany miners' produce back to Adelaide, where a better price for their gold than in Melbourne was guaranteed, helped ensure that South Australia survived. Henry was pleased to report to Graham that the colony was returning to something like normal, with excitement about gold subsiding. Walters and Williams were now both off the scene, and mine and smelting works had hung on. While also improving his own position as a man of property and, increasingly, a financier, Henry's hard-headed management of the SAMA's business had helped the enterprise, and thus the largely copper-dependent colony, to weather some very stormy times.

6

A CUCKOO IN THE NEST?

New Year 1855 marked a strong upward turn in Henry's fortunes. Life in Adelaide had returned to a normality not seen for four years, before the gold chaos. He had two pieces of good news for Graham. The water in the mine had been 'forked' – pumped dry – and, though labour was not cheap, because the cost of provisions was high, he expected the mine to be in full swing again within six months. The three-year drowning of the underground workings had produced little serious damage, and the forking was achieved with no accidents to 'men, machinery, or mine'. Coaxed back to life, 'Schneider's Engine' was 'in famous working order'. Loyal Captain Roach got a vote of thanks after the AGM, but not a bean of a bonus. Graham, though, could expect a dividend payment come March which, Henry reassured him, would be paid 'promptly'. It was time to retrieve the horses from pasture, invite Henry's roving eye, Captain Bryant, back from Bendigo, and sound out the colonial government on getting Burra's police force back up to strength. Things were looking so promising that, in a few months, it was time to ask the SAMA's London agents if they could get something into the City articles of *The Times*.

The agreement with the E&ACC was due for renegotiation, and Henry lamented the lack of any real competition in the colony for reducing the Burra's produce. Despite the E&ACC's stronger negotiating position, Henry still believed that no company could make the Kooringa works pay, given the logistical problems and attendant costs. He asked Graham to surreptitiously find out from 'Doctor' Schneider what the E&ACC's plans might be, without revealing their own position.

On the personal front, Henry recorded: '13th July 1855 Took possession of House Acre "30" North Terrace, Adelaide.' With this address, the Ayers moved up in Adelaide's social world, into a cottage owned by SAMA Director William Paxton. The property now known as Ayers House, comprising a museum of the Ayers family and a privately run restaurant, was then a more modest-sized dwelling. Lease it was all Henry could do for many years, even after Paxton returned to Britain later that year, as Paxton refused to sell. This did not prevent Henry's making extensive improvements. He transformed it, probably with architect Kingston's assistance, into a more spacious, and gracious, building which befitted his rise in the social scale.

From this new home, it was a comfortable walk to work. Virtually chained to a desk, Henry valued a good walk, book-ending his days with a constitutional. Residents of various suburbs, if they were 'up betimes', might encounter him taking in the eucalyptus-laden air. After writing all day, he might slip out for a further two or three miles before taking up his pen once again. Lucy Lockett Ayers believed that he even won a running race, though his own note that, at a Queen's Birthday 'Pic-nic' at the William Elder property Ridge Park, in 1861, 'HA' won a running race is more likely a reference to his son Harry.

Late in 1855, with a more pleasant residence in a prominent location and, seemingly, a more secure future with the mine fully operational again, Henry diverted himself with some lighter social activities, such as concert- and theatre-going. These particular performances were clearly highlights for Henry, as he kept a specific note of them. A wild fortnight saw world-renowned exotic performer Lola Montez stir Adelaide audiences. She and her top-quality supporting troupe performed – extremely well – nothing more titillating at first than conventional comedies and farces. Then she informed her audience that, after a few nights, she would be performing her celebrated Spanish Spider Dance, replete with castanets. Naturally, the Victoria Theatre was packed to the rafters, though she apologized as she came breathlessly back onto the stage for giving a cut-down version of the dance, having been indisposed all day. If they would only patronize the next performance, she would give full rein, acting out the effects of a supposed tarantula bite. 'Did it please you?' she panted to the house, and the captivated patrons chorused back 'Yes, yes!' at the tops of their voices.

Fed the dance in instalments, no one was incensed by its sensuousness. Madame Montez had been led to believe that the dance which pleased fastidious tastes in Europe and America would not be accepted in Adelaide, but her faith in her public there was not misplaced. Her farewell performance

– not complete without her leading man's grotesque performance of the song 'Villikins and his Dinah', which had the whole house in stitches – coincided with the arrival of the news of the fall of Sebastopol. The colony could breathe a sigh of relief. The Russians were done for in the Crimea. Perhaps la Montez had been just what the staid men of Adelaide had needed, in the second year of the Crimean War, with its attendant problems for trade and commerce.

A little more decorum graced the performances of the Hungarian violinist, Miska Hauser, who next swept into town. Accompanied by a Berlin pianist, and with a soprano as light relief, Hauser charmed his audiences, again including Henry, particularly with his own fantasias on themes from popular operas, and also with similar on Irish and Scottish national songs. He provided the finest musical performances Adelaide had enjoyed since the visit of world-touring soprano Catherine Hayes, a year before; Henry had not missed her performance either. The popular King William Street medical man, Dr Kent, had been able to do the Irish nightingale a service before her performances in Melbourne, where he found himself in time to restore her lost voice.

Miss Hayes' duetting protégé, a Monsieur Coulon, had unfortunately left her, in Melbourne, when she would not increase his fee. To give her a breather, two Adelaide gentlemen supported her local performance. Alderman Lazar, who had been a 'professional Thespian', volunteered an aria from Rossini's *Cenerentola*. The Musical Director also helped out, with a couple of comic numbers. The second half opened with the *Barber of Seville* overture and then Catherine Hayes brought tears to the eyes of Scots and Irish present, with performances of their countries' ballads, and to the rest of the house with a performance of 'Home, sweet home'.

Theatre-going, to which Henry had been no stranger in his younger days in Portsmouth, was a pursuit to which he was also introducing his children. Frank Ayers, at just 12, wrote a delightful letter to his school-master Richardson Reid, enthusiastically telling of a theatre outing on which 'Dadda' had taken him and '[his] brother Harry'. The Nelson Family had performed a vaudeville at the Exchange Rooms based on a tale of a young man's search for a silent woman for a wife. The trip had obviously made a big impression on schoolboy Frank, who pours out the whole story willy-nilly to 'Dear Sir'. The unkind farce involved the young man uncomfortably declaring his love via an ear-trumpet to a prospective fiancée feigning deafness – and dumbness. His shouting renders him 'dumb' while she 'recovers' and scolds him, revealing her ploy. Frank signs off, 'Your dutiful Pupil' in a well-formed hand. At the same age, his father was already out in the wide world of Blyth's law-office.

Political Diversions

A different form of entertainment which Henry may have enjoyed was the second election to the South Australian Legislative Council, a little earlier in 1855. The new members would be charged with the task of developing not just a 'representative' constitution, as had been in force since 1851, but a constitution for 'responsible' government. Henry's name is just one of the couple of thousand appended to a petition to the Queen for a new Constitution, but not the one so far on offer. A big public meeting resulted in this petition which was fervently against an upper house of members nominated by the Governor and, what was more, for life. That was not South Australia's way. The colony of Victoria was getting an elected upper house, and did its neighbour deserve less? Nearly 20 years after its foundation, South Australia would take complete responsibility for its own governance, so long as the decisions of its parliament were not in direct conflict with the interests of the Mother Country and the Imperial parliament in Westminster. As with the election of 1851, this Legislative Council would include a number of close associates of Ayers including Peacock, Kingston, Fisher and Younghusband. Henry was on much more than nodding acquaintance with all these influential gentlemen.

Fresh Faces

Just before Christmas, Henry wrote to John Benjamin and Louisa Graham that her young brothers, Henry and Frank Rymill, had safely arrived aboard the *Caucasian*. With no immediate employment in Adelaide, they were experiencing country life upon a Captain Hughes' Yorke Peninsula sheep station. Henry noted, with amusement, that the attractions of station life had already worn thin. As it happened, Ayers was able to offer a position in the city office of the SAMA to the elder boy, Henry Rymill.

With work piling up, Ayers needed more support in the city office, and he needed a back-up. He gained the SAMA Board's support for a reshuffle, with MacDonald to assist him more closely, and a new clerk to be taken on. In the New Year, Henry Rymill shook off the burrs and dust of the country and escaped the blazing summer sun, settling for an office berth with the SAMA. He was to be initiated into the line of Cashier, under Henry Ayers' eye, at a promising starting salary of £150 per annum. Young Henry, then, would soon be established in the office routine of arriving by 9.30 and staying until all work was completed, with no overtime. There was no lunch break, each munching a biscuit at his desk. Younger brother Frank remained sweltering on Hughes' pastoral property. Ayers remarked that they both appeared good chaps and reassured Louisa, and the boys' mother, that the Ayers would be kind to them; they had already become 'great pets' with the Ayers children.

Over the next few months, the mine approached normal production levels, and the E&ACC also strove to reach its full potential, with an influx of new Welsh smelters. All was not sweetness and light at the smelting works, though. New and old men could not agree, their fractiousness aggravated by the Christmas/New Year revelries. Still, the smelting facilities were reaching completion, with a second furnace house, powered by the draught of an 80-odd-feet stack, swinging into action, after the gold hiatus. Stubbornness may have been the predominant trait of both the existing workers and their animal counterparts; the South American mules were increasing in numbers. The beasts were sufficiently willing, though, to ensure a more reliable supply of fuel. The dark question then arose in Ayers' mind, not so much whether the smelting works could cope with the ore raised, but whether the mine could – long term – produce sufficient ore for the furnaces.

Ayers was always ready to respond to calls for specimen ores from various institutions, whether the Crystal Palace, Edinburgh University, or the museums of Copenhagen and Cape Town. Even the newly opened Burra Institute – with a generous donation from the SAMA of £200, plus £30 for books and £20 towards the first year's expenses, and 'warmest wishes' – got its collection. It was as well to keep in with big names too, including Baron Rothschild, in Paris. The big-name institution that did not want to know was the British Museum. Ayers understood that South Australia was underrepresented among its collections. Months later, news arrived that the Museum was quite happy as it was, thank you, and had passed the box of specimens on to the School of Mines.

The E&ACC's use of mules would prove a comparatively short-term expedient because, before long, a railway edged northward, Henry selling some of the Prospect House grounds for the track. He also warned Graham that a dividend was unlikely at year's end due to the sizeable fall in the London copper price. Louisa's brother Frank had also been able to give up the country life. Following a brief stint in an Adelaide wine merchant's, he had a position in the office of the Police Commissioner, at £125 a year. Both the Rymill boys were in the house as he wrote. Nor was Ayers neglecting the wider Graham family, letting Graham know that he had already written to the Adams. He was pleased that they had settled well, which he had gathered from a long, cheerful letter from the couple.

By November 1856, Ayers was glad to hear that the Grahams had had another son, Frederick Malcolm, and reported that, once again, the Ayers were expecting. During the period for which there is no surviving correspondence between Ayers and Graham, the families had both produced a child apiece. The Ayers' Arthur Ernest had been born towards the end of 1852, while the Grahams had had a daughter the previous year. In mid-December, at the age of 44, Anne Ayers bore Lucy Josephine, possibly named after cousin Lucy Lockett Baker, who had been in the colony for some three years by then. At any rate, she became Ayers' 'Josey'.

The Baker family was in the forefront of the Ayers' minds at this time for, just previously, Ayers noted a tragedy in his *Memorabilia*: 'Arthur Baker Drowned in River Torrens'. Anne Ayers' teenage nephew was swept away while crossing the swirling river at the Frome Road ford. It was a 'black spot'. Arthur Baker, despite the efforts of three policemen to rescue him, and of several doctors to resuscitate him, could not be revived.

By the Christmas of 1856, the Ayers' house rang to the voices of six surviving children. With this growing weight of family responsibility, Ayers became concerned that the future for him, the Grahams, and so many others, might not be particularly rosy. Their main support, the Burra mine, was again not looking too promising, with lower-quality ores the deeper the mine went. There was no trading in SAMA shares on the Exchange in Adelaide and Henry felt he should state it formally that Graham's income from his Burra investments was likely to diminish.

The Books Cooked

Henry Rymill's developing abilities in the city office may have helped to uncover fraud. In mid-1856, Ayers had the unhappy task of writing to the wife of one of his 'all good and true' office staff. In addition to his sympathy for Mrs MacDonald's predicament, with her husband too ill to continue work, Ayers offered six months' salary for the distressed couple. Unfortunately, MacDonald turned out neither good nor true. The salary offer was swiftly withdrawn. It seemed that MacDonald had squirreled away over £1,600 by fictitious entries and alterations to documents, and an 'extensive system' of forgeries. An Extraordinary Meeting of the Directors was hastily called. From the minutes, there appear to have been no recriminations with Henry about how this could have happened, presumably for some time, under his nose, and those of auditors *and* the directors who rotated in forming Finance Committees. Perhaps no one's face felt totally free of egg.

It would not have been so bad, either, if Henry had not already made a very bold public claim, in front of the annual Estimates Committee of the legislature: 'We have such a state of accounts [in the SAMA office] that, without egotism, I am satisfied there are no books in the Colony kept better than ours.'

Oh dear! With what he believed was a tight regime, with Cashier MacDonald's actions checked by Clerk Joseph Phillips as a first line of defence, how could the deception have been so successful? MacDonald spent most of the day at the counter. His duty was to pay, for example, the draymen, which was a trickier and more responsible task than that performed by any bank teller: his day was a round of rapid calculations of what was due; he had no vouchers to work from, just experience. From early on, Henry must have had considerable faith in him, as he had been one of those to whom Henry had been prepared to lend funds – £300 – once he had acquired the spare cash. Clearly, MacDonald had enjoyed reasonable room for manoeuvre and had used it. What may have started in a small way, perhaps, had got out of control, playing on his mind, and very likely undermining the cashier's health.

SAMA counsel E.C. Gwynne advised taking no action, for the moment, while the sinking MacDonald was in the Asylum – 'lunatic' as he was described. Replacement cashier, Phillips, would move up to assist the Secretary, while a lucky Henry Rymill won speedy promotion to full cashier, and a salary increase to £200. On this touchy subject of office irregularity, Ayers' letters to Graham maintained a discreet silence.

From then on, the board stipulated that Ayers was to countersign all cheques and Bills of Exchange. For good measure, too, Association officers were to give securities worth twice their annual salaries – a touch daunting, one would have thought, for a young man such as Rymill, recently arrived in the colony and fallen into a well-paying job. The Secretary was prepared to personally back him, putting up £200, and a similar guarantee soon arrived from the young man's father in Britain. Airily, Henry had dismissed the giving of securities as unnecessary when queried by the Estimates Committee six months before.

The younger Henry's abilities were further acknowledged by the time of Ayers' May letter to Graham of 1857. There they were, virtually knee to knee, working at the same table. By then, Ayers felt confident in suggesting that some of his responsibilities could be passed to the young man. He was 'always ready and willing with helping hands' and, Ayers said, he knew no one who had taken more interest in the mine nor, indeed, in Ayers' private matters. Perhaps the 36-year-old Ayers was seeing a younger version of himself being formed in his image and under his guidance – the two Henrys, coincidentally, shared a birthday and Henry Rymill's initial training ground, before emigration, had also been a law office. True, the youth had a lot to learn yet, but 'if his health and life were spared', his mentor believed he would do well. Doubtless for Louisa's reassurance, Ayers adds that her other brother, Frank, was also giving satisfaction in his office. The boys were certainly finding their own feet, for Ayers reported that they had decided upon a cottage of their own. Ayers assured the Grahams that this was at their own instigation. He could never get them to complain of their old quarters – in the Ayers household. He hoped they would not 'indulge themselves too soon' but, as he had noted their 'steady habits', he was not too worried.

Rocky Moments with Graham

Just as Ayers was writing this commendatory letter, J.B. Graham, in Germany, was penning a letter of profound dissatisfaction. Its tone could easily have broken their business arrangement, if not permanently damaged their friendship. He was seriously put out, feeling that Ayers was not keeping him sufficiently up to date about his South Australian interests; someone in London reportedly had more of a grasp of recent developments than he had, in Frankfurt. He was sorry to have to write in this vein but, if Ayers could no longer find time to keep him fully apprised of developments, he would have to make other arrangements.

This unpleasant news travelled fast and Ayers was soon replying in most hurt tones. He protested that he had always carried out Graham's instructions to the letter. Itemising all dividend payments since 1854, he pointed out that only two had been a fortnight late, and *that* because the SAMA itself had caused the postponement. What was more, Ayers pointed out that, when he forwarded payments, he gave Graham favourable treatment by paying at half a per cent lower than the bank's commission rate.

Shortly after this rocky moment, another of Ayers' long-term friendships underwent a transformation, with the repatriation of James Coke, whose name joined a lengthening list of South Australians heading Home. It is not clear why he went but it is possible that he had recently been widowed; evidence is scant. So many familiar faces had departed, but Henry, shouldering ever wider responsibilities, could not afford the time for even that longed-for visit Home.

By September 1857, it became clear that Graham had heard nothing from Ayers after his letter of complaint. Henry could only write and say that he had replied immediately. As if to further appease Graham, Henry reported that, as he had received no further instructions about the recent dividend, worth some £400, he had been successful in investing it locally for Graham, at the handsome interest of 12 per cent. He could procure a similar return for any further funds at the same rate, or at least 10 per cent. He also had some encouraging news of a find of good-quality malachite ore, which quite reminded him of old times. Unfortunately, the water now encountered in the deeper workings was 'very quick' and the SAMA was sending to England for a yet more powerful pumping engine. Generally speaking, things were going very well in South

Australia, where favourable conditions had promoted good crops. In fact, the colony was looking as prosperous as he had ever seen it.

In closing his September letter, Henry referred to the recent 'discourse' between Graham and himself, hoping that the air was now sufficiently cleared to allow them to return to normal dealings. A stiffish opening to the letter gave way to a much chattier conclusion, including the fact that the Rymill brothers were engaged at that moment in tomfoolery with the Ayers children. In a postscript, Henry asked after the Adams; he had written them several letters with no reply. Ayers must have been writing purely socially, having no business responsibilities for them. It seems unlikely that Graham's disgruntlement would have rubbed off onto his stepfather and caused a distinct cooling in what had appeared quite a warm relationship between the Adams and the Ayers. Perhaps it was just a matter of 'out of sight, out of mind' from the Adams' point of view – they had been back in Britain for a good two and a half years by then. Maybe advancing years or infirmity played a part. Whether Henry's persistence in writing to them was from genuine warmth of feeling, or to keep up the relationship with Graham through Adams, is impossible to tell.

The Burra's prospects, and share prices, see-sawed, and Henry hoped Graham's European interests might be compensating for poorer returns in South Australia. He informed Graham that he had made a donation on his behalf of £10 towards the building of the Burra Institute – a sum he himself matched. The provision of this intellectual and cultural centre for the raw mining town was in keeping with a movement for workers' institutes which had begun in Adelaide's satellite village of Hindmarsh in 1847. Adelaide's own Institute had opened later that same year with 280-odd members; according to the *Register*'s report, this was not quite matched by the number of books on its shelves. The SAMA's original £20 grant towards the running costs of the Burra Institute had not been intended as a regular thing, as Henry had stressed, yet a second and a third year's grants were forthcoming. While the SAMA was laying out its initial donation with one hand, it was soon clawing back sums from the community with the other; Henry began calling in arrears of rent built up over the gold period. He and the board had at least given some time for the community to get back on its feet before sending demands. The mine was again a good provider of employment for the community, but how long the enterprise would remain bountiful was uncertain.

7

'... INTO PARLIAMENT HE SHALL GO!'
(*Iolanthe*, W.S. Gilbert)

The year 1857 had held some disappointments for Henry Ayers. A long and mutually lucrative friendship had come within an ace of foundering, another long-term friendship had changed through repatriation, and the financial prospects looked less bright the deeper the Burra mine was dug. However, a wider opportunity for Ayers had opened up in the form of a political career. This should not be construed, though, as an opportunity to advance his financial prospects; at least, certainly not directly. At that time, there was no pecuniary reward for members of the new South Australian parliament, as indeed was the situation with the Westminster original. To be elected was a form of public service, undertaken in what leisure time your business interests permitted. True, it would be a boys' club where you could rub shoulders with other influential colonists, but there was no money to be directly made, unless you aspired to be one of the five ministers forming a government.

It was considered that it would be only men of affairs, those with landed interests, or those with experience of the world of business or the law who would be able to devote the necessary time to the legislature. Clergy were specifically excluded. In any case, only those sorts of men – no women, of course – would have enough interests tied up in the smooth running of public affairs and therefore the necessary responsibilities inducing them to govern judiciously for the good of all. That was the theory, anyway.

The new constitutions given the nod by the Westminster parliament in 1856, providing responsible government in the Australian colonies, also provided for the possibility of occasional meetings of a Colonial Assembly. The Cabinet in London must have thought this very sensible – varying customs duties and, very practically, varied gauges for the railways could be avoided. However, South Australia's sense of itself as a special case of a colony meant this found little favour locally. Another objection was that it would be difficult to persuade men of affairs to neglect their own interests for periods of weeks, which is what attending an intercolonial conference would entail. It was difficult for South Australian representatives to attend, say, in Sydney, and was asking the earth of any potential Western Australian representatives – if or when that colony got responsible government. Before the election campaign was over, though, the *Register*'s editor was prepared to assert that there could be advantages in federation, even seeing it as 'inevitable' in the long run.

The members of the South Australian legislature of 1855, who had thrashed out the form of the new Constitution for their colony, led the world in democratising their institution. They were well in advance of Westminster by giving every male colonist the right to vote, with no property qualification whatsoever for members of the lower house, the House of Assembly. This incidentally also applied to the Aboriginal community, which had been endowed with all the rights and responsibilities of the British citizen, with no choice offered in the matter, when the Province was proclaimed 20 years earlier. Not even Gladstone, in the Third Reform Bill in 1884, gave the entire male population of Britain the vote. The last unenfranchised quarter would have to wait until the end of the First World War, at the same time as Britain moved to give the vote to women over 30.

As big a question for the drafters of the South Australian Constitution had been whether there should be an upper house. The men of interest wanted some kind of brake upon the legislation passed by an Assembly elected by universal male suffrage, so an upper chamber was deemed desirable. Some even wanted to ape Westminster closely by reflecting the hereditary principle of the House of Lords. Particularly, 'Pioneer' colonists like John Morphett fancied some kind of title running in the family, perpetuating the influence and authority he felt pioneer colonists had, and were due. In this, however, he was wandering far from the principles of the abandonment of an 'establishment', for which he and people like him had supposedly pioneered South Australia in the first place. His baronial aspirations were howled down, mocked in a cartoon in *The Monthly Almanac and Illustrated Commentator*, for example, as 'Ye Lorde offe Cummines' – an allusion to Morphett's house, Cummins, to the city's south-west. There would be elections for an upper house, the Legislative Council, the 'brake' being achieved by setting the bar both for standing and voting rights fairly high, in terms of property qualification.

In a number of editorials during early 1857, the editor of the *Register* debated the pros and cons of a bi-cameral parliament. He feared the upper house's becoming a repository for 'political incapables' and past-it politicians. Conversely, the Legislative Council might be peopled by 'men of ability, political knowledge and character', to the detriment of the quality of the lower house where the power, especially to initiate money bills, was vested. How could people of the right ability be induced to stand with no offices to attract them, nor patronage and, certainly, no money?

The only person forthright enough to announce his candidacy, as early as the New Year, was Marshall MacDermott, the SAMA's former banker, a magistrate and erstwhile friend of Lord Byron. No younger men came forward to join the spry 66 year old. The *Register* editor was exercised about what damage would be done to South Australia's wider reputation if it failed to live up to its self-devised Constitution. Candidates who held off until the writs were issued, on 2 February, would be required to remain silent about their policy stances, except in print. The newspaperman speculated that this would allow the interviewee, should he choose, to 'misunderstand' the question put to him and, instead, to 'answer one' he 'wish[ed]' he 'had been asked'. The world of politics changes little. It is nice to note in passing that the affectionate/derogatory term 'pollies' for politicians was in general use in that era, as it is still in Australia today.

As the fourth week of 1857 began, the editor was finally able to name a handful of candidates joining the lonely MacDermott, including John Morphett – barony or no – William Younghusband and James Hurtle Fisher, outgoing Speaker of the superseded Legislative Council. The total number, including those yet to accept their nominations, came to just 17 candidates, who would still be able to play a couple of rounds of musical chairs before anyone would be short of a seat when the music stopped.

His newspaper awash with lengthy reports of political meetings, the editor felt it necessary to castigate the meeting-goers, at least in the city, for rowdy behaviour which would have shocked a recent arrival from, say, London or Edinburgh. In London, he maintained, the speakers would have been received with respect and their arguments listened to with thought and courtesy; a public meeting should, he maintained, consist of speakers who came to speak and the public whose part was to listen. South Australian meetings were discreditable in that everyone came to chip in his six-penn'orth.

The question was raised about the lengthy term that a Legislative Councillor could be expected to enjoy, namely 12 years, whereas the Members of the House of Assembly would have to face their constituents in a mere three. Especially as it appeared that the membership of the upper house would largely be inexperienced, and possibly silent upon their policies before election, it was suggested that a member should be willing to resign on demand from the electorate. The editor's suggestion – actually a provision of the Constitution – was a roll of the foundation members, in order decided by lot, with the first third retiring after four years, the second third after eight, and only the final third sitting for the full term. It was thought unlikely that members would stand for re-election after such a lengthy service – the editor seems very naive here.

New Blood

In this same edition, a letter of nomination was published from voters of the Burra and Clare district. About 160 names were appended in support, 'respectfully request[ing] that you will allow yourself to be put in nomination as a Candidate for the representation of the colony in the Legislative Council'. The addressee was 'Mr. Henry Ayers, Esq., Adelaide'. The drafters continued: 'Your eminent fitness for the office, and long connection with this district, induce us to seek the honor of nominating you as a candidate.'

How did the good electors of the Burra and Clare district come to the conclusion that Henry should be their man? Perhaps Paddy Kingston, standing again for Burra in the lower house, had put the suggestion about during his attendances for the frequent public meetings in town. If there looked like being a dearth of candidates, it is quite possible that Kingston cast about for suitable possibilities. He had known, and worked closely with, Henry for the previous dozen years and was in a good position to judge his potential for the role of parliamentarian. It would have taken no effort to drop a word in the right ears around Burra and then it would have been up to those in the district to make what they would of it. Ayers had certainly spent enough time in the vicinity of the mine, over the years, for the populace in general, as well as the miners themselves, to have formed an opinion of his capabilities. He would certainly have been known to them as a no-nonsense, hard bargainer, who generally meant what he said and stuck to it. He would not have been a sentimental favourite, that is certain.

Whose would have been the right ears to bend? Not prominently placed among the 160 names were those of Captain Roach and two others of the Burra mine's captains. The name of the mine Accountant, William Challoner, also appeared, though, interestingly, that of Assayer Elphick did not. Another noticeable name was that of William Woolacott, a timber carter for the E&ACC. Ian Auhl (1986) retells the description of him riding into Burra atop the large boiler cylinder of a big pumping engine he was transporting, waving his topper to the assembled townsfolk. All these named individuals, then, had a vested interest in keeping in with Henry, if Kingston had dropped a hint. Alternatively, as they had had numerous opportunities themselves to gauge Henry's mettle, they may have independently concluded he would be a good candidate, and it would still do their own situations no harm to appear in the list. A considerable percentage of the rest must have comprised Burra miners. Even if men could now make their political choices in the privacy of the election booth using the secret ballot, it still did not hurt to be seen supporting the company's influential Secretary.

The newspaper printed the nominee's response:

> It affords me great pleasure to accede to your request, and allow myself to be nominated as a Candidate for the representation of the colony in the Legislative Council, the more so as the solicitation comes from those who have had the best opportunity of judging of my past conduct.

Now, whether by accident or design, it looked as though Henry Ayers was headed for a parliamentary career. He was not the sort of prospective candidate that the *Register*'s editor feared might hold back his name until it was too late to scrutinize his policies. Below his acceptance of nomination, Henry clearly set out his stance on the important issues of the day:

> Inasmuch, however, as [my?] connection with the political offices of the province has not been of a very prominent character, I deem it due to you, and to the electors generally, to state, briefly, my opinions on those subjects which now engage public attention.

He dealt with the most important issue – that of 'Waste Lands' – first. He supported the status quo, avoiding the risk of a 'market value' undermining the investments of South Australia's contemporary landholders; he favoured an 'up-set price' for saleable land, as had been South Australia's method since its inception. The question of immigration next received his attention. He favoured an annual regulation of numbers, via the level of funding from lands purchased, tailored to suit the colony's labour requirements. The purchasers of that land should, he felt, have first call in nominating prospective immigrants. Those fortunate to be accepted for immigration, and supported by the government's revenues from land sales, should not be able to quit the colony within a certain period, unless able to repay their subsidized passages. One wonders whether he had any twinge of conscience about the nature of his own assisted passage, 17 years earlier.

The next topic concerned the possible abolition of duties on the distillation of spirits by private individuals. Henry's ability to deal with a hot-potato issue like a consummate politician is revealed in this policy – or non-policy – statement. He hedges, deftly maintaining that the matter is too important to decide upon 'without a full enquiry and very careful consideration'. He had once had to speedily learn the politics of the copper world, and that ability was now clearly in evidence in this new sphere. On taxation generally, he went on, he was confident there was no need to raise any new taxes, unless the disappearance of distillery duties required a replacement from some other source. If they did, then he felt it should be done by the whole community shouldering the cost so that there would be no extra burden on trade.

Rail, he knew, was an important policy area to canvass and Henry shows a typically hard-headed approach. Though he had expressed almost envy, to J.B. Graham, about the luxuries and ease of travel which the returned South Australians were enjoying in Europe, Henry was of the opinion that his own stamping ground was nowhere near ready for any great expansion of its modest railway network. He wanted no further revenue raised to fund government-built railways. The only proviso he made was that, if any experimental track were to be laid for a trial of horse-traction – and there was considerable debate about the comparative merits of horse and steam – then that track should be capable of supporting steam-traction thereafter. The two could then be evaluated against the alternative option of merely macadamizing roads.

To round off his platform, Henry confirmed his opposition to state borrowing unless marked benefits could be shown. Improvements of amenity for the citizenry he felt should be funded out of surplus revenues. On the possibility of the state's becoming involved in providing general education, he only observed that, if this were to happen, it should be entirely secular in nature. To conclude, Henry wrote:

> My only motive in seeking the honor of a seat in the Legislative Council is to serve my country faithfully and to the best of my ability, and that I have no interest apart from the permanent good of South Australia. I unhesitatingly pledge myself to do so,
>
> And am, Gentlemen,
>   your obedient servant,
>   Henry Ayers,
> North-terrace, Adelaide, January 21, 1857

Whatever ulterior motives Henry may or may not have harboured for wishing to be an actor on the new parliamentary stage, this sonorous note of commitment to his adopted land begins to sound convincing. With the passing of the years and the more settled nature of the colony post the gold era, Henry was finding himself ever more entrenched in the colonial society in which he had chanced his future. Gone were the days of longing for the management of a SAMA London office, with the possible advantages it might have afforded Henry and his family. All that now remained of that sentiment was the desperate desire to at least pay a visit Home, though that possibility seemed as distant as ever, with a parliamentary involvement now further complicating matters.

Twenty-seven candidates eventually accepted nomination for the 18-seat upper house. *Register* editor Anthony Forster, in bemoaning the quality of some of the candidates, never thought it necessary to draw his readers' attention to his close personal interest in these matters, beyond including his name in a list of upper house candidates. Forster, like John Baker, Morphett and Fisher, had sat in the previous version of the Legislative Council.

On 23 February, a remarkably orderly gathering took place on the Corporation Acre on King William Street, an acre set aside by Colonel Light, in his 1837 survey of the city of Adelaide, for the use of a future city council. A booth was erected for the official nomination of upper house candidates. By the editor's estimate, a hundred or so people either took part in or watched proceedings, with as little excitement as might have greeted the election of a 'parish beadle'. The booth had to stand at least a hundred yards from the nearest public house, to comply with the new regulations, so there was none of the drunkenness, debauchery, taunting and scuffling which the editor associated with electoral activities, generally. Henry's name was listed second, after George Fife Angas', the man who, next to Wakefield, could be considered to have done the most to bring South Australia into being.

Henry needed an official proposer and a seconder. The name of neither was prominent in the community, unlike numbers of the other proposers. It is interesting to see Paddy Kingston's name crop up several times as a proposer and seconder of candidates, but his name was not associated on this occasion with Ayers'. Henry was the official proposer of William Younghusband, underlining the friendship between the families. The editor conceded that even those with no political experience could at least bring in new blood.

Despite the fact that some of the candidates seem to have expended a fair amount of money keeping up their public profiles with, in some cases, daily notices in the paper, Henry appears to have chanced all upon his one statement in accepting nomination. However, it can have done his image no harm, a day or two before the poll, to have his name on the front page announcing the twenty-ninth dividend for SAMA shareholders; there is nothing like being associated with success.

The ninth of March was polling day. We can only imagine the excitement and apprehension in the Ayers household – Henry does not seem to have put in an appearance in the SAMA office for a few days on either side of election day. All seems to have gone smoothly in the first election involving adult male suffrage – at least for the lower house – and the secret ballot, two of the points that the Chartists had demanded in Britain 20 years before. A nod was also given in the direction of a third Chartist point – short parliaments – given the House of Assembly's three-year terms.

With polling centres stretching across many hundreds of miles from Adelaide, with a pretty primitive range of transport options, it would demand much patience before upper house results could be declared, but by 18 March, things seemed clear. Henry's nominee, Younghusband, was in, and so was his one-time employer, Fisher. The eminent lawyer, and SAMA solicitor, E.C. Gwynne was also in the chamber, as was the SAMA's agent at the Port, Captain Scott, and the title-less John Morphett. If there was nothing else for the Legislative Council to do, the *Register* editor opined, at least a codification of the colony's statutes could be usefully undertaken utilising the considerable legal experience of both Gwynne and Fisher. On the twenty-fifth, the final list of new upper house members showed Henry coming in at number 13. At not yet 36, he was the youngest member of the upper house – the newest new blood.

Taking His Seat

Henry was sworn in on the first sitting day, 22 April. A President had to be elected for the upper house. James Hurtle Fisher slipped back into his still-warm seat as he had been the previous 'Speaker' of the Legislative Council under the old Constitution. With that formality over, the members of the lower house – with their newly chosen Speaker, man-of-many-parts George Strickland 'Paddy' Kingston – were invited to attend in the Council's chamber for the inaugural Speech from the Governor, Sir Richard MacDonnell. A guard of honour from the army Regulars was drawn up outside the recently extended Legislative Council chamber, with the Mounted Police also in attendance. Sir Richard rode the short distance down the hill from Government House, with some officers forming a procession with him. That was about the extent of the panoply the colony could muster.

As the *Register* commented, this was to be the first opening speech in the colony's history not written by a Governor. Rather, it was the work of the elected representatives of the people; for the moment, though, the ministry led by B.T. Finniss, which had been in power under the old Constitution, had been allowed to carry over as a sort of caretaker government in the new parliament. A Reply was agreed and standing orders devised, which required Legislative Council Members to sit on Tuesdays, Wednesdays and Thursdays from one o'clock, initially, in the afternoon. The first question set down on the notice paper in Henry's name was on a topic close to his heart. He wished to know what money was sitting in the fund from the sale of lands that should be directed towards supporting further immigration; one can safely surmise that he was looking for an increase in the number of miners brought to the colony to reduce the resistance of his mine workforce to lower wages. The first committee he was put on was to look into the comparative costs of running the city-to-port railway, and the Gawler Town line, by animal traction, as opposed to the existing locomotive power. Captain Bagot, the owner of the Kapunda mine, and a fellow Legislative Councillor, also proposed the extension of the Gawler line – surprise, surprise – to Kapunda. He favoured horse-traction, for cheapness, and to get the railway up and running quickly.

In this first Session, though, Ayers' name is not prominent in putting down questions on the notice-paper, but the large majority of the upper house membership's names do not appear at all. Known to be able to count, Henry was often called upon to act as Teller for the 'Ayes' or the 'Noes'.

One has a sense, too, that some of these founding members chose to play at being parliamentarians. Modelling their procedures largely on the precedents of the Westminster Parliament, virtually without thinking, they failed, for instance, to grasp an early opportunity to improve on its efficiency by sticking too closely to precedent. At one point, the upper house declined to accept a 'Message' from the Lower – that is, a bill which had been passed by the lower house which was then carried to the Upper for consideration – because the lower house was not actually sitting at that precise moment. The argument ran that the British House of Lords customarily accepted messages from the House of Commons when both the Lord Chancellor was upon the Woolsack and the Speaker was in his chair.

Similarly, the summoning of the Assembly Members to the Legislative Council for the Governor's Speech opening the Session was just a formula drawn from the ancient precedent where the monarch had a throne in the Lords, among his or her peers. There, the lesser fry in the Commons could be summoned to attend. In the South Australian parliament, by contrast, it was somewhat anomalous that the Members of the House of Assembly, who had been voted there by the adult male suffrage of the entire colony, were – if not 'summoned' – 'invited' to attend in the upper house, whose members had been elected by a narrower clique of the slightly better-to-do. As the Queen's representative, the Governor could hardly be considered to be among his peers there. It was quite clear, though, that the royal stand-in was seen to be most closely aligned with the influential 'interests' found seated in the upper house.

Modernization was in the air, and a major step forward for the colony took place in the October of 1857 with the inaugural train journey on the new Adelaide to Gawler Town railway. It is doubtful that freshman parliamentarian Ayers would have missed the occasion. With the push and press of the crowds, it is a wonder that the train got away reasonably on time – a mere 20 minutes late, due to the tardy arrival of the Port Adelaide train. Punctuality itself, apparently, according to the editor of the *Register*, in comparison with British railways then – and of more recent times! The Adelaide–Gawler railway could boast a natty line in rolling stock, too, with freshly varnished and gilded carriages festooned in flags and bunting. The longest train ever to run in the colony until then had a 'push-me-pull-you' start, with an engine at the rear helping to overcome inertia. The only real blot on this maiden run concerned a little accidental flag-burning: a spark from the engine – itself garlanded with flowers – set a Union flag atop a carriage alight. An intrepid passenger managed to get the blazing duster down, despite the train's running at speed, and successfully doused the flames. A radically minded wag suggested that, if they could not keep a Union Jack aloft without accident, perhaps it was time the colony became independent and flew its own flag.

Election to the new parliament added to Henry Ayers' workload during 1857, but his involvement in South Australian politics provided a new outlet for his abilities. Perhaps it mitigated some of the mounting disappointments regarding the prospects of the commercial concern nearest his heart, the Burra mine.

8

ARRIVALS AND DEPARTURES

Late in February 1858, in her forty-sixth year, Anne Ayers was delivered of her last child, a brother to Frank, Harry, Fred, Maggie, Arthur Ernest and Josey. The arrival of little Sydney Breaks, his middle name perpetuating Henry's mother's maiden name, may have necessitated house extensions at North Terrace; new bedrooms had just been added and a library begun.

Six weeks later, at nine in the morning, Ayers was first to greet, in friendly fashion, another Adelaide arrival. Ayers was doubtless delighted that, after a ten-year absence, J.B. Graham had finally responded to his frequent urging to 'pop out' for a visit. To Graham, his friend and agent appeared 'very much smaller' than he had expected. Had Ayers' central importance in the management of Graham's South Australian business interests caused Graham to develop an inflated image of Henry's stature?

Graham arrived alone. His wife, Louisa, was doubtless anxious to have a report on her brothers, overseas now for more than two years, but no motive was powerful enough to impel her to accompany her husband, perhaps fearing that it might encourage him to resettle. Graham himself may simply have been curious about changes in the colony, though perhaps he was especially keen to check up first hand on Henry's handling of his affairs.

Together with his 'shrunken' friend, Graham made immediately for the SAMA office where he found brother-in-law Henry Rymill at work, looking 'quite jolly'. The magnate also paid a quick call on Frank Rymill before driving around the town, which Graham found 'much improved'. In addition to government buildings on Victoria Square, the classical façade of the Supreme Court building would have been new to him. The old Legislative Council building, on North Terrace, modified for the use of the new parliament, would also have been unfamiliar. Graham took a peek at the Hindley Street shop where he had worked prior to sinking his entire savings so successfully into the Burra mine, finding the premises empty and shabby.

The following day, Graham attended St John's church with Ayers and his three eldest sons. All the boys, he noted, seemed to like Louisa's brothers. Old faces and new public buildings pleased Graham and soon he was back in the old routine, attending a SAMA Board meeting, where there was encouraging news from the mine. By the time he had taken in days at the races and lunches at the 'club', whichever that may have been, he must have felt quite at home again.

After about five days, though, it was time to get down to a serious discussion with Ayers about the future of his business interests in the colony. Their arrangement for Ayers' agency was altered: whereas Ayers had initially accepted a fee of £120 per annum and no commission, he would now accept £100, with 1 per cent on mortgage interest and 5 per cent on rents collected.

Graham spent a great deal of time with the Ayers. The following day he drove with Henry and Anne in their carriage and pair to Holdfast Bay and, come the Sunday, he attended St John's again, prior to dinner at the Ayers' house. The afternoon was spent in the Botanic Gardens, across North Terrace from the house, which Graham felt would be 'a great boon to the town'. Enjoying a 'glorious' sunset from Ayers' garden rounded off the day, though not before he saw all over the house.

A visit to Port Adelaide, while the steamer for Melbourne was in, caused him to note the improvements there, then he and Ayers enjoyed an evening of whist. The card games must have been lubricated with some of the fruit of the vine because, the following morning, Graham was 'not quite jolly' and Ayers 'not quite the thing'. A morning of inactivity was self-prescribed, though whether Ayers also had the luxury of a long lie-in on that sultry morning is doubtful.

The SAMA AGM had a small turnout and for some strange reason, considering Graham was just visiting, they re-elected him a director, another board member having agreed to step aside. Graham seemed pleased with an outcome which gave him some element of control, 'ex officio'. Little over a month later, however, an Extraordinary General Meeting was called to re-elect his predecessor. Still, Graham had been happy with his brief time in the limelight.

Graham was up early next morning for a jaunt to the Mid North with Ayers and Harry, in a chaise and tandem, taking in the 'much improved' town of Gawler. After a day and a half of travel, they reached the newer SAMA mine, Karkulto, Graham noting as they went that the countryside was 'sadly' in need of rain. The party was met by the Burra mine's Captain Roach, whom Graham perceived as looking 'quite well and fat'. When Graham last saw him, Roach was just a subordinate officer under the licentious Superintendent Burr, but for nearly the entire length of Graham's absence, Roach had been in sole local command of the Burra mining activities. Graham presented a silver cup to him as a token of his esteem.

The next day being Sunday, Graham attended morning service, finding the singing 'very good' – praise indeed for the Burra Anglican community from someone who had revelled in the choral sounds of Liverpool Cathedral and York Minster. After lunching with Captain Roach and the Ayers, father and son, they inspected the neighbouring Bon Accord Mine – fortunately not a SAMA investment. Graham observed that it did 'not look very jolly'.

The following morning Graham, Henry and Harry Ayers 'donned [the] miner's rough flannel shirt, felt hat, moleskin trousers and coat and went down to the "50" [fathom level]' of the Monster. The sight of magnificent malachite ore, in various greens, gratified them all. Graham echoed Ayers' impatience to see the major pumping engine – Morphett's Engine – working so they could explore what the '60' would hold. They spent a couple of days snooping around the whole of the SAMA's area, admiring the Assayer's garden and his collection of valuable specimens. The mine's complement of a hundred horses was inspected as well as the town's slaughterhouse, a facility Henry had had to insist upon for reasons of community health. Once the other directors arrived, deliberations could take place regarding the mine's future, and a couple more evenings passed with a 'rubber of whist'.

A circus in town made the place quite gay, Graham observed. Bread and circuses were all that the townspeople were likely to enjoy, because a major outcome of the directors' visit was a wage cut. This was one of the last occasions on which they would still have the whip hand as far as their workers were concerned, with few significant mines competing for their skills.

At the end of April, they made south-westward, towards Clare, where the SAMA had bought some mineral land. A lucrative visit followed to an 800-acre sheep property of E.B. 'Paddy' Gleeson. He had dealt with Henry for some years, for example, maintaining some of the mine's horses on his property during the gold period. Now the 'King of Clare' needed a loan, which Graham provided at a handsome 12.5 per cent interest rate. The homeward-bound party joined a train leaving Gawler Town for the city, where they were met at the station by the Rymill brothers. Letters from Louisa and the Graham children were waiting for Graham, and he was still enjoying re-reading them the following afternoon, at 'the foot of the waterfall', with Ayers and Henry Rymill, whom he thought 'made for life'.

The magnate threw a 'champaigne' lunch at the SAMA office for Ayers, the Rymills and the clerks, before friends and family saw him off at the Port. They all had bundles of mail for despatch by the *Admella*, and then the Graham visit was all over. He sailed at four in a 'squall of wind', having to anchor no further out than the lightship, where he felt 'very queer and sick'.

The last few days of Graham's visit could not have been solely a gay old social whirl for Ayers, because he had prepared for his departing friend a letter outlining all the Burra grandee's South Australian interests, Ayers' views on them, and the new arrangements for Ayers' fees. He notes Graham's total South Australian wealth at comfortably over £73,000. As ever, Ayers addresses his friend formally but rounds off with the hope: 'long may we be spared to correspond to our mutual advantage'.

The Last of the 'Old Ones'

The two men had lived in each other's pockets and surely reinvigorated their friendship. The closeness between the Ayers and the Rymill boys must also have helped to cement relations. In forwarding mail from Louisa in June, Ayers cheerfully reiterated his feelings: 'everything satisfactory' and they would 'in future mutually assist each other'. Henry added, 'For gossip see my letter to Mrs. Graham.'

Henry maintained a fairly optimistic tone in mid-year, with the copper price and share prices reasonably high. One or two darker clouds appeared, though, with a local downturn in Adelaide, the firm of Marks and Gollin having already gone bust. In parliament, Henry had unsuccessfully striven to get immigration increased. 'Better luck next time,' he quipped to Graham. His idea was, doubtless, to procure an influx of miners so that he could again erode wages as costs were rising with deeper workings. There was a heavy weight of work upon him: 'You must excuse me. I have been writing since dinner at a rail-way rate and it is now past ten o'clock. Good night! My fingers ache – .'

Towards the end of 1858, though, Henry was living it up. He spent only two evenings at home in a fortnight, having been 'very dissipated' at dinner parties, balls and concerts. Now he must make up for it, with innumerable letters to write and the half-year Report to prepare – 'Well! You know that I can work, and work.' Voyages were on his mind, with Ellis, Thomas Waterhouse and Peacock, all closely bound up with the Burra concern from its inception, embarking for London. A feeling of being left high and dry once again is evident when Henry remarks: 'I am the last of the old ones! Positively without a contemporary. "Well what's the odds so long as you are happy!"' This final overemphasis seems to belie the words – he definitely protests too much.

Ayers warned that the best times looked unlikely ever to return, likening their three mines – Burra, Pompurne and Karkulto – to a family of brothers, 'where one boy works to keep the three'. He raised one optimistic note: Captain Bryant, the Burra's Second Captain, had been sent hundreds of miles northward to look at some mineral lands there, which the SAMA was thinking of leasing.

Henry was grateful to Graham for advice on striking the most profitable balance between selling copper in London and locally in Adelaide. He also warned his friend not to leave himself too bare of funds in South Australia – it seems Graham had wished to repatriate some of his capital. Henry joked that he would not want Graham to require to borrow off his friend as he would find Henry's terms pretty stiff – in fact, an 'unprofitable transaction'. It was obviously in Henry's best interests to induce Graham to keep as much as possible of his capital locally, which Henry would manage for him, now that he was partially reliant on commissions.

A late Christmas present for Graham was that Captain Roach had put aside 17 hundredweights of choice malachite. When it was made up to a ton, Henry would see it sent over, together with some wine and other local produce Graham had ordered. The choicest specimen of the malachite no doubt wound up in Graham's admired marble and malachite table exhibited at the 1867 Paris Exhibition, its green matching the envy in the eye of Queen Victoria's son-in-law, the Prussian Crown Prince.

A Sad Affair

It would have surprised the Burra magnate when the next official letter found him, in Frankfurt, that it was not in Henry Ayers' hand but that of Henry Rymill. Having dealt with the business matters, Graham's brother-in-law continued:

> a little local news part of which I have to communicate is sad for only yesterday at 3 o'clock p.m. Mr. Ayers lost his dear little Baby which had been suffering for a long time from diarrhoea and last week the malady appeared to have left the infant but alas! So weak that he could not rally. I daresay you will remember it when you were here it looked such a fine healthy child – they had named it at our suggestion Sydney – although so young the occurrence has caused a sad gloom at the Terrace – it is the second.

The death of Sydney Breaks Ayers, just short of his first birthday, was a devastating blow to Ayers, adding to the previous loss of little Charles Coke. Within a couple of days of Sydney's death, the elder Rymill brother was keeping his brother-in-law updated on the aftermath of the domestic tragedy in a succession of letters: 'Mr Ayers will not be able to write to you this Mail owing to the sad affair, before mentioned.'

Ayers was apparently away from the office at least through the first half of February, when the baby was ill, as Joseph Phillips, the Assistant Secretary, was handling correspondence. It seems the grieving father attempted to go into work on the fifteenth – the day after little Sydney's funeral – managing to get through dealing with just one letter before having to put all the other matters back in Phillips' hands.

A month later, bringing home the full effect of the loss, Henry Rymill wrote saying that Henry and Anne, together with their eldest, Frank, had gone to Melbourne on a

> pleasure trip . . . he having been so so in health since the death of the infant and I was afraid at one period that the sad bereavement would have been the cause of completely knocking him up and therefore I strongly urged upon him the necessity of his getting away from home for a short time and which I am happy to say he soon concurred in – packed up and was off in two or three days afterwards leaving the whole of his business matters in my sole hands, with power to effect loans, receive monies etc., correspond, open all letters private or otherwise, to draw cheques ad libitum, . . . [illegible] me to reside at his house (during his absence) on North Terrace (where I am now addressing you from) and generally to do as he would if he were here, – What do you think of this? . . . I have no doubt . . . [unclear; that?] he has tested what I was like, during his trips to the Burra and found me <u>up to the mark</u>.

Of course, this month-long absence was a different kettle of fish from the odd days away in Burra for which Henry Rymill had previously covered. Young Henry, though, was 'delighted to say', that things had 'gone remarkable well and it shall not be my fault if it does not end so and if it should I think there will be more likelihood of his some morning popping across to England'. There is that nonchalant use of the word 'popping' again, as though voyaging to the other side of the world were no more than a day-trip.

It seems as though Ayers was very much on the edge of a precipice, staring at emotional collapse. The consummate professional, who had never taken any significant time off from his SAMA duties in 14 years, was forced to accede to pressure to get away; the show could simply not go on. Henry Rymill was, of course, anxious to prove himself in his mentor's absence.

By mid-April, Ayers was sufficiently recovered to take up his business interests again. He recounted for Graham the recent unhappy circumstances and his visit to Victoria. The family had visited Ballarat to get a first-hand look at the gold-fields. Maybe with some sense of satisfaction, born of inter-colonial rivalry, Ayers reported that the gold-fields were not what they were. Henry had been surprised to see an 'old Burra-ite', who treated the Ayers 'very kindly' and with 'some affection'. Henry seems surprised and touched. Had this unexpected affection come from someone whom Henry had given little cause to have shown him any?

Ayers reported how well Henry Rymill had managed affairs during his absence. With no good news of the two newer mines, Ayers told Graham of a giant gold nugget they had seen in Victoria; it took two strong men to lift it and it sold for £9,325 – Graham would certainly have bought it, Ayers enthused. There was one piece of encouraging news: Captain Bryant gave a recommendation in favour of the Far North mineral lands acquired by the Chambers family. The SAMA was thinking of trying to persuade the Chambers to sell, although Ayers was not optimistic of agreeing reasonable terms. This proved the case, and the SAMA bought some adjoining sections instead.

The debt over the Prospect House organ was finally paid off by Christchurch, though only under the threat of litigation, Ayers told Graham. He would have gone through with it, too, he laconically added. He had recently heard that the Grahams' little boy's health had been in a 'dangerous state' and hoped for better news, adding that he and Mrs Ayers could sympathize with the feelings of the Grahams. He was relieved to hear, soon, that the boy had recovered.

The Burra roller-coaster was ascending again by mid-1859. So long as the price was right, Henry felt a well-managed company might have a chance at making something of the new land near the Chambers'. If the subject were mooted in London, Henry warned Graham not to let on from whom he had got his intelligence – as though there would have been much doubt.

Pleased as Rymill was with his own performance in deputizing for his mentor in the latter's private business matters, he would have been minding his Ps and Qs as SAMA Cashier. The previous year, in the wake of his powerful brother-in-law's visit, young Henry had obviously grown a little too big for his boots, treading uncomfortably on Assistant Secretary Joseph Phillips' toes. Beyond a complaint of insubordination, there was clearly some reference to favouritism on the part of Secretary Ayers – if not a kind of 'nepotism'. Phillips wrote directly to the board, putting the members on the spot. They obviously valued Phillips and did not want to lose his experience, particularly in the aftermath of the MacDonald business. On the other hand, they could not allow him to impugn their trusty Secretary's name. They gave him a week to consider whether he might withdraw his letter, and apologize to Ayers, or whether he would insist on having his complaints officially investigated by the directors. Phillips chose the former course. To ensure office harmony during any further absences of the Secretary, Ayers was required to write down the lines of demarcation between the responsibilities of Cashier and Assistant Secretary. If Phillips had already had to swallow his pride, henceforth Henry Rymill would have to swallow his. Any disputes were to be resolved by subordinating Rymill's views to those of the Assistant Secretary.

Appeasing the Magnate

Less than a year after J.B. Graham's visit, despite all the relationship building, he was clearly disgruntled again about Ayers' service, though his main complaint seemed to be a shortage of family news and gossip. Henry used the heat as an excuse, since the southern January and February were not months when 'men willingly entail labor on themselves'. However, both disputed letters were, he continued, 'friendly and business like. We have not much to say about ourselves [the Ayers]. We lead a very quiet life particularly at the season of the year I have just alluded to.' To make a point, he also told Graham that he was using 40 minutes before the closing of the Mail to write this extra letter. Mollifying Graham further, Henry promised, much as one might attempt to get round a peevish child, that his next letter would be a '<u>very very long</u> one'.

Henry Rymill had also just written to Graham, commenting on Graham's visit to Italy on the return journey from Australia. Ayers had immediately stated his intention of viewing the same historic sites, such as Pompeii, though Rymill thought it unlikely, as he had cried wolf so many times. However, if he once made up his mind, thought his young acolyte, 'many' would 'be sorry to see him go'. It is doubtful that young Henry would have been among them because he added an immediate reference to the success of his deputizing while Ayers was in Melbourne.

Henry Rymill went on to give full insight into Ayers' methods as a financier. He first informed Graham that the Australian Trust, with British capital, had set up an Adelaide branch, which would probably make borrowing easier and cheaper. Of course, this would not be a happy prospect for the capitalist likes of Graham and Ayers. Whereas Graham's funds could usually net him, under Ayers' watchful eye, far higher interest rates than he could get for his European-based capital, the arrival of a new lender in town was likely to see progressively lower returns.

Not only was Ayers highly successful at achieving good returns, through having 'such a connection', but the security of the investments was also first class, Rymill believed. He gave the example of the purchase by the Halletts of their leased sheep run, north of Burra. It was to be auctioned and the lessees required some support for their hopes of purchasing. Ayers would lend them money at the rate of 15 shillings per acre; whatever price the land sold for above that, they would have to find themselves. The land would be purchased in Ayers' name and he did not require a mortgage deed, which was a practice the Australian Trust, for example, was restricted from matching. The entire loan was of the order of £2,500 over three years, at a rate of 11 per cent. The Halletts were content with the arrangement, given Ayers' reputation, knowing they would then have the land conveyed to them on repayment of the debt. Cheerily, Henry Rymill observed: 'it is the very best security it is possible to have'.

In Ayers' July 1859 letter to Graham, Ayers revealed another money-making ploy: the SAMA share price had risen to £146, following news of a good new lode. If it reached £150, Ayers would sell some of his own and Graham's shares, then hold the funds until the price dropped again, when he would repurchase. Ayers, of course, had inside information that would assist in the timing of such transactions. Adding to the likely profit, Ayers would invest the held funds in the meantime to gain some interest. As he occasionally did, usually when rendering account to Graham, he congratulated the magnate on what he [Ayers] had been able to achieve for him. He was particularly pleased at this point because he had been able to lend out Graham's funds, for mortgages, for good rates despite the plentiful supply of credit locally. With things going so well, Ayers had managed to get a long lease on Paxton's cottage on North Terrace, with major alterations planned, including a 'Large Drawing Room', or Ballroom, to the 'Eastward'. Bolstering the relationship, Ayers added that, although there was little time for gossipy detail, his boys' school holiday would finish shortly and they and their friends were having a 'sort of a revel', with 'Master Frank Rymill in his delight at the head of them acting charades and all kinds of larking'.

Ironically, Ayers takes a passing swipe at the SAMA directors for their greed; while on his recuperative visit to Melbourne, he could have sold 200 tons of copper there for £100 a ton – a slight premium on the London price. The directors, though, wanted to hang out for £103. Ayers also spoke of the £1,000 of Graham's funds which he had remitted to him as requested that month, despite unfavourable bank terms. Ayers was thus emphasizing his following of his friend's instructions, although, as the man on the spot, and with Power of Attorney regarding Graham's South Australian share dealing, he could see that it was not to Graham's best advantage.

Before year's end, though, Henry again raised the question of his responsibilities as a proactive agent, especially given the distance and time taken for instructions to arrive. Making yet more placatory noises, Ayers hoped his care for his client's 'safety' did not expose him to any risk of offending Graham. After all, he felt that his own local knowledge should be employed in ways which would benefit his friend. There must have been a certain element, too, of trying to minimize Graham's exposure through his over-optimistic expectations of the Burra's future. The reality was, as Ayers highlighted, that the costs of working the deep levels were not compatible with paying dividends and the shallower levels had already given up much of their bounty. As for the mineral lands acquired in the Far North, if they were worked, they would have the double-edged effect of lowering the price of copper and increasing the demand for labour, which would only increase costs for the Burra and the other SAMA properties.

The Unforgiving Sea

Ayers' reluctance to write to Graham about non-business matters was overridden by the 'greatest calamity that ever befell the Province' – the loss of the steamer *Admella*, on which both Graham and the Ayers had themselves recently sailed. Nearly all the 'seventy souls' on board, mostly South Australians, were lost. The ship had plied the Adelaide–Melbourne run for a couple of years without mishap. 'I am truly sorry for Captain McEwan who was a careful and kind man.' On a routine voyage, in good weather, she had foundered on a reef off Cape Northumberland just before dawn on 6 August. She had split in three and it was very difficult for any in the forward sections to reach comparative shelter in the stern. Survivors of the initial shock certainly needed what protection they could find because it took over a week to rescue the few who proved able to withstand cold, wet, hunger and the temptation to drink sea water. Three lifeboats were lost or damaged, and it took the bravery of two seamen, paddling for an hour using bits of the ship's fittings on a makeshift raft, to reach the surf-pounded shore. The alerted Cape Northumberland light-keeper took a fall from a frisky horse, and the news eventually got to the electric telegraph station in Mount Gambier by the efforts of the young son of the horse's owner.

Adelaide was thrown into turmoil, with frequent telegraph reports updating the anxious populace like a macabre serialized story. It took an entire week of failed, or aborted, rescue attempts by small boats before a handful of survivors was got to safety. One unlucky man, having made it into the open boat, drowned within yards of the beach when the frail barque capsized in the surf and the exhausted Captain McEwan had lacked the strength to hold on to him.

James Hurtle Fisher, 'our old acquaintance', lost his son, George, in the first moments of disaster, though son Hurtle survived. Fisher had had to step down from the President's Chair in the Legislative Council due to his level of anxiety on the afternoon the news of the wreck reached Adelaide. His son-in-law, John Morphett, deputized. A Mr Humble – no doubt S.W. Humble, a James Coke predecessor at the mine's Store – had had a 'very wonderful' escape, which seems to have been in the form of a last-minute decision not to board the doomed vessel. Graham cannot imagine, comments Ayers, the intensity of concern in Adelaide at the receipt of each telegram, arriving three times a day for a week. Business stopped and, at one time, it was believed all were lost.

To close his dramatic letter, Ayers soothes any feelings of neglect Graham might feel, once again, with the promise that his next letter will be a 'long epistle'. Both Rymill brothers were assisting Ayers in meeting the deadline for the Mail – this seems a fairly regular occurrence and, in remarking on it, one wonders whether Ayers was trying to underline the closeness of collaboration with the young men, or suggesting that they were themselves going beyond the call of duty. Frank Rymill, after all, had no duty beyond the call of friendship.

The Frailty of Existence

In September 1859, Henry Rymill informed his brother-in-law that Ayers was pleased with him and that he and his brother, Frank, had bought some land on East Terrace. He hoped that, when Graham next visited, he would stay at 'our house'. Before the year's end, Graham was being smothered with offers of comfortable accommodation; the Ayers were up to their eyes in bricks and mortar and so would be in a position to offer Graham 'a little breathing room'.

Both Henrys were in for a little Christmas treat: for the first time, the SAMA's city office would be closed. From lunchtime on Christmas Eve, the staff could relax over the festive season, not showing up for work again until the twenty-ninth.

The Rymills had concerning family news, though. Their brother, John, a sheep farmer 400 miles up the River Murray, had been forced to seek medical advice in Melbourne after a bout of rheumatic fever. It would not be his only brush with illness and, when he later appeared in Adelaide – a gregarious, accomplished bushman as Ayers thought – his city-dwelling brother, Henry, rather looked down on him. Fondness for drink worsened John's health problems; he could have 'made a mark' in Australia, possessed as he was of 'intelligence, ability and gentlemanly demeanour'. Meanwhile, Ayers condoled with the Grahams over the death of another Rymill brother at Home. As if to reassure Louisa Graham regarding her Adelaide brothers, he described them as going along in their usual 'jogtrott' way.

With middle age approaching, Ayers was well-established in his home, social and business life. He was working hard to maintain a good relationship with the SAMA's major shareholder, whose brother-in-law he was grooming as a potential successor. He now had the added challenge of a parliamentary career. Yet the loss of little Sydney shattered the professional façade, and other events, such as the wreck of the *Admella*, seemed to reinforce the fragility of life. Was time slipping away, and with it the opportunity to see Home again?

9

'BETSEY' NO LONGER AS FREE WITH HER FAVOURS

The Rymill brothers' planned substantial East Terrace house would firmly establish them in Adelaide society. At the start of 1860, Ayers 'cheerfully' organized an interest-free loan on Graham's behalf, considering them both worthy and 'very deserving young men'. The project may well have derived from the redevelopment of the Ayers' own house. The flurry of drawings, tenders and contracts during the most recent upgrade – particularly the addition of the ballroom – must have stimulated the young brothers' aspirations. The Ayers' facilities to entertain were being greatly enhanced, what with the 'folding doors' that, actually, slid away allowing the ballroom to expand into three areas adjoining.

The mine was becoming a headache, though, the deeper it was dug. Good finds were becoming rarer, causing a falling share value. Matters were made worse by a declining world copper price. The other two mines in the portfolio, Karkulto and Pompurne, were one moment hopeful, the next, hopeless. Ayers saw new ore-dressing machines being trialled at Burra, a mechanization designed to reduce labour costs. He had already cut the men's wages, admitting that it had not been popular but, with emphasis, told Graham it '<u>must be done</u>'. A dry winter meant high-priced provisions for miners' families, and he just hoped that a good harvest would alleviate pressure upon them. That was a forlorn hope, however.

Now it was Ayers' turn to be in a huff with Graham, who would not accept his agent's advice. Maybe he should not bother giving his opinion in future, Ayers grumbled, and simply follow Graham's instructions. The opinion of the man on the spot clearly went for nothing. Graham tried to glean independent advice from an increasingly experienced Henry Rymill who, wisely, counselled heeding Ayers, with Graham's investments 'very snug' under his agent's care.

By 1860, the Burra faced competition. Not only were some men drawn to gold on the Snowy River but, more seriously, the Wallaroo and Moonta mines opened on South Australia's Yorke Peninsula. With the world copper price dropping further, Ayers' and his associates' SAMA shares were now worth half what they had been at the dizzy heights of £200-plus. Ayers was prepared to concede that the Wallaroo – its name a contraction of the local Aboriginal words for wallaby's piss – was generally considered good but would not be worth what had been paid for it. It obviously rankled that some felt the Burra was as nothing to this new mine with '9 miles of copper, etc, etc.' If that were so, he told Graham, it had to be 'a wonderful place'. Amid these developments, one month Graham would have Ayers buy shares, the next, he should sell. It was impossible to keep up. Next, Graham wanted a massive withdrawal of his capital, under Ayers' highly successful management, from the colony, which would mean a sizeable reduction in Ayers' commission.

Ayers' knowledge of mining was called upon by a parliamentary Select Committee looking into the potential of mines in the Far North, where the SAMA had dabbled a little. Questioning was generally courteous but sometimes deeply probing. Ayers coped calmly, never stuck for an answer, sometimes with a light touch of ironic wit; it was a coolly professional performance. This was in marked contrast with mine-owner Jamie Chambers – more comfortable on the box of a coach than in the witness box. Ayers' level-headed analysis sought to save the colony's good name at Home. To accomplish this, he made careful distinction between what information had been given out by the South Australian Government and how this had been represented by the British press. Typically, too, he exhibited his preference for local investment for developments, be they rail or government-supported road haulage, so that profits would not disappear overseas. He also displayed his keenness for saving something for the future, an attitude apparent earlier after Roach had reported some new-found lode. It should simply be recorded, Ayers had felt, telling Graham it would come in handy one day for Graham's son. '[A]s Captain Cuttle says "When found make a note on't"' had been his comment – a Dickens quotation printed as a motto on the flyleaf of one of Ayers' scrapbooks.

Despite suffering a 'relaxed sore throat', Ayers had to make another appearance before a lower house Select Committee, this time on Building and

Friendly Societies, some of which had run into financial difficulties. From his financial experience generally, and his positions as Trustee of three Adelaide Building Societies and director – or 'Trustee' – of the Savings Bank, he made several observations and recommendations. His own idealistic belief in self-reliance, typical of a self-made man, came in for gently sceptical criticism from the Committee members. Onerous regulation on directors, he believed, would make it more difficult to get good men: 'A good man often doubts his ability, when a man less scrupulous thinks himself well adapted for the post.' Enactments would not make bad men good. As in the Northern mines inquiries, his testimony was fulsome and frank; he saw no virtue whatsoever in the position of Trusteeship in the Building Societies, and would have withdrawn from those he held if it would not have entailed added costs for them. He and others like him, in giving their names to various societies, were only lending an aura of probity; this was not actually a fraud, but little short of one, he believed. To avoid actual mismanagement of funds, he recommended the introduction of passbooks for clients, to be regularly checked off against the offices' ledgers. That simple innovation, plus a thorough checking of the books every six weeks as was the practice in the SAMA office, would also pick up any discrepancies early. His holier-than-thou testimony must have been based with some confidence on the post-MacDonald embezzlement reforms in his own office. Some were calling for government regulation, with the back-up of an auditor, but Ayers saw little benefit. Letting all shareholders be responsible for probity would be the best form of regulation. If government regulation required auditors, then the fastest, least careful firms would be preferred by the societies, so the object of the exercise would not be achieved. When governments thought they were closing off loopholes, Ayers asserted, 'clever dishonest men look about for some weak point to assail, and it is singular how successful they are'.

Generational Change

Henry Ayers may have bitterly repented not making his trip Home. New Year 1861 brought news that his mother, Elizabeth, had died in November, at the age of 77. Regretfully, he had to admit that he had long cherished the wish of seeing his mother once more, and of showing her some of his children – 'a sad disappointment to me'. He had to bear it with what resignation he could muster – 'It has been willed otherwise. I must submit.' She was buried in the cemetery in Deadman's Lane, associated with St Mary's, Portsea. It had been just over 20 years since Henry had quit Home and family.

The only trip Ayers could manage was a holiday to the other colonies, with 'Mama', Harry and Maggie, again testing Henry Rymill's potential for holding the business fort. While the cat was away, the mouse purchased a further 11 shares for Graham. Even if young Rymill was in tune with his brother-in-law's wishes, one wonders how pleased Ayers was upon his return.

Henry Rymill finally plucked up courage and proposed to Anne Ayers' niece, Lucy Lockett Baker – a young lady Graham had noted as pretty during his visit three years before. Rymill had, as he himself reported it to Ayers, 'gone and done it'. According to Ayers, the young couple appeared very fond of each other and, as he laconically put it, it would be 'their own faults if they are not happy'. Fortune was certainly smiling upon the young couple. The Nova Scotian stepbrother of Anne Ayers, W.H. Lockett, had died, leaving £750 for Lucy, £600 of which was to come to her immediately. This might have proven a godsend as, with Frank Rymill moving out of the East Terrace house, the responsibility for its upkeep would now rest with young Henry alone. It would also be his sole responsibility to pay off his brother-in-law Graham's £1,000 loan by the following January, and Ayers entered a plea for Graham's consideration in this matter.

In a very wet month, the weather turned out perfect and the young couple married in Holy Trinity Church, North Terrace, with Maggie Ayers as a bridesmaid. The breakfast, back at the Bakers' house, was attended by the Dean, the Postmaster-General, and 'other celebrities', which included the Rymills' early employer W.W. Hughes of the Wallaroo mine – one wonders if he would have been quite as ready to attend had he known that the groom had recently reported in none too glowing terms, to Ayers and Graham, on his mine's potential. According to young Henry, Ayers made a 'particularly strong' speech, from which modesty prevented the young man quoting. The bride left for East Terrace 'amid a perfect shower of old boots and slippers'. Later, a ball at the Ayers', organized by Anne, was a 'perfect success'.

The young couple started their life together with undoubted advantages. They had a fine house and furnishings – 'everything that hearts can wish for'. Ayers pointed out to Henry and Lucy that they started married life in as happy circumstances as most people would achieve after a lifetime's toil. At 40, Ayers was sounding quite the paterfamilias. It had been decided that the gooseberry, Frank, would get a house on his own where he would be looked after by 'old Abigail' – doubtless, a suitably matronly housekeeper previously to the two brothers.

More social activity ensued, with the two Franks, Ayers and Rymill, having a successful sporting trip to Melbourne; as members of the Volunteers, they were competitive shots. Nor was there any shortage of entertainment for Easter, 1861, as the Bianchis' opera and pantomime company was in town. Ayers attended their performances, though there is no record of whether he availed himself of the other acts on offer: the dancing Leopold Brothers with Fräulein Fanny, 'Australia's Fat Boy', a satirical elocutionist and three enormous salt-water crocodiles.

Frank Ayers' playing part-time soldiers was one thing, but what sort of career did his father have mapped out for him? Once Frank completed his studies at St Peter's College, he was articled to the firm of Fenn and Bruce, by then the SAMA's lawyers. Ayers joked to Graham that he had 'arranged to prepare the new firm [a reorganized SAMA] by making one of the members "learned in the law"'.

Managing an Ageing Betsey

Still puzzled over the potential of the Peninsula mines, Ayers found it difficult to concede that Wallaroo was rising while his dear 'Betsey Burra' was showing signs of age. The actual mine output fell to about a quarter, and a trial of a new process for poorer ores – the Henderson Process – was postponed on grounds of cost. International events also clouded the horizon for South Australia's produce as the 'American difficulties', as Ayers called the Civil War, would throw their shadow over world trade. To ensure the Burra at least broke even, Ayers was forced to make a hundred miners redundant. Like a surgeon removing a limb to save the patient, Ayers felt

> There was no help but this. . . . A reduced establishment and diminished expenses, or certain loss was staring us in the face . . . It was a disagreeable business, but the safety of the concern demanded it, and we did our duty.

Ayers may have assuaged his conscience with his belief that those still working were earning weekly what their Cornish brethren earned in a month. After the Half-Year General Meeting, in the October of 1863, the *Register*'s editor could even then talk in terms of those, labouring in England, who might still 'stare to see the good wages' being got at the old Burra, despite the 'wonder of the Peninsula mines' – Kingston's description.

News of gold discoveries in New Zealand had thrown the workforces of the neighbouring colonies into a frenzy but, so far, the Burra was unaffected. The colonial government also resumed supported immigration, something Ayers had always favoured. He had written to the colonial administration in 1847 proposing the intake of Chinese or European labourers and 'artizans' from Singapore. Coming from Portsmouth, it is quite likely that Ayers was used to, and fairly accepting of, a diversity of nationalities and peoples. An icon of Britishness like HMS *Victory*, for instance, had 20 nationalities aboard her at

Trafalgar. He was not as 'modern' or 'multicultural' in his attitude, however, as to be proposing a permanent incorporation of an Asiatic population; he had proposed one-year contracts.

The position in 1861 was different, though. The Chamber of Commerce, particularly, was petitioning for the lifting of charges on Chinese men entering South Australia. Others opposed this, with thinly veiled racism amid fear that the colony's British heritage would be diluted. The overt complaint was that the Chinese only came to amass wealth and then take it home. What, one might ask, had Graham, Paxton, Beck and their like done? Ayers tabled a petition in favour of his pet project – supported immigration from Europe – and seconded the Chief Secretary's motion proposing the ending of the impost on Chinese immigrants. The motion was presented in a spirit of doing unto the Chinese 'that which we insist they do unto us'; theirs was a reform, they claimed, done on principle, not for expediency.

In Need of Cheer

As Christmas 1861 approached, the business prospects of Ayers and Graham looked very unappealing. Captain Roach measured a lode in inches instead of fathoms. Startled, Ayers wrote to check that it was not a slip of the pen. Ayers surmised that Graham would be spending a great deal of 1862 in England. If only he, too, could be there. Soberly, his assessment was 'I will have to keep to my post here, sometime longer yet I find.'

The summer proved extremely hot. Writing to his sister, Louisa, Henry Rymill described a far-below-par Ayers:

> Mr. Ayers I am sorry to say has been suffering a good deal from indigestion and in low spirits, we do all we can to cheer him up and I think he will soon be enjoying his usual good health – if he does not get better soon I believe it is his intention to take a trip to Queensland with Mr. ['Billy'] Sanders for a change of air.

Unfortunately, Ayers was unable to gain relief from the testing South Australian summer through a sea-voyage. Quite the contrary; he had to visit the inland Mid North on an inspection of the Burra mine. 'I will not forget it easily' – it was 'during a regular hot week and I got nearly roasted'. By way of compensation, at least he could report that the '70' was so much better than the '60'.

Henry Ayers was in need of some cheer just then as things generally were in poor shape in South Australia, with numbers of Adelaide businesses failing, and a disappointing harvest. Ayers had to apologize to Graham for quite the worst Annual Report that he had ever had to put out. If it had not been for the cost-cutting of the previous year, they would have been operating at a loss.

William Henderson – the 'Process' man – was proving awkward. Graham had obviously thought that a £10,000 trial of the Henderson Process was a mere 'trifle' but Ayers, cautious as ever, said he had to know it would work before sanctioning one – rather obviating a trial, one would have thought. Once again, Ayers and the SAMA directors proved themselves timid and risk-averse. Ayers felt it would be a simple thing for either the man himself, or a representative, to come out to assess the Burra's potential for its implementation. It struck Ayers that the obvious candidate would be the E&ACC's Thomas Williams, who had had seven years' local experience as Superintendent of the Kooringa smelter.

The June letter of Ayers to Graham reads almost as a sort of 'Credo' in the context of his utter commitment to the Burra mine and its success over the years. He had demanded, and would demand, sacrifices from all concerned with it – except perhaps its shareholders – but at some personal sacrifice as well. With the copper price locally down at £88 a ton, the Burra's shares were correspondingly down at a £100; he would buy more for Graham but it was a sad reflection on how things looked that the Burra's shares were now down below the level of other South Australian mining ventures. He had received a long letter from Thomas Williams, which made reference to some misgivings on the part of Gregory Seale Walters about the application of the Henderson Process in the Burra context. So, Ayers' old sparring partner from the initial days of the Patent Copper Company was still involved, to a certain extent, in the Burra's future.

Ayers had to admit that, if he had got the best of Walters and Williams in the past, they seemed 'disposed to get the advantage of me this time [in connection with Henderson's process] . . . Mr. Williams offers us much less than we are getting from the E.& A.C.C. I want more!' He added, 'I am ready to meet Henderson's people fairly liberally and honestly. But I mean to make money out of them. I do not care what they make out of us.'

Graham may have noted that Ayers' earlier comment, that he was prepared to deal 'liberally' with Henderson's people, had now moderated to 'fairly liberally' – and Ayers had obviously felt a need to add 'and honestly'. He may well have been concerned that he had given Williams, and Walters, cause to feel that he was just as slippery as the aptly named Seale Walters himself. The two referred to Ayers as being 'sharp and unbusinesslike' over calling for a reply from them by return, which he refuted.

One of the points over which he was at odds with Williams was that the latter did not believe it necessary to continue washing the ores, with Henderson's

Process. Ayers declared that the secret of success with South Australian mines was to work the ores up to a high grade – through the various sorting, jigging and laundering processes, before handing it over to smelters, or whomever. The other South Australian mine owners, he believed, had not yet learned that lesson.

On a brighter note, the young Mrs Rymill had been confined with a daughter – 'Mother and child doing well', Ayers reported to Graham. A month later, both Mrs and Miss Rymill were ready to make an appearance in society, with a celebration for 140 held at the Ayers' for the Rymills' first anniversary. Louisa Graham was aunt and Anne Ayers great-aunt to the same little Rymill infant, Florence Edith. Henry Rymill had been in the fortunate position, as well, of becoming a capitalist, able to afford the purchase of two Burra shares, with the price having fallen to £90. Graham might have looked askance at hearing that news from Ayers. The new father had written to his brother-in-law obviously in a state of shock on hearing Graham's terms for the extension of his loan for the East Terrace house. Henry's conception of what any 'small rate of interest' should be, and only then if his brother-in-law really desired it, obviously did not accord with Graham's ideas. The latter had clearly asked for a far steeper sum and the demand hurt. Happiness was also tempered within the Rymill family with news of the death of father Robert Rymill.

A Ray of Sunlight

For Ayers, though, September 1862 saw the clouds begin to part and he could even look forward to making that return visit Home in a little over a year. A novel project on the horizon concerned the pastoral properties of two of the surviving Fisher brothers, who were in financial difficulty. What a happy position Ayers must have felt himself to be in to be able to organize a rescue package for these Pioneer colonists, whose 20-odd years of striving looked like ending in failure. It must have been a prospect to savour, for an assisted emigrant like Ayers, to have reached the position of being able to bail out the sons of as prominent a South Australian as Sir James Hurtle Fisher. The two men sat in the Legislative Council where Fisher's Presidency brought the knighthood. In addition to drawing Graham and two others into the scheme, Ayers planned to put in £30,000 of his own money. Once this business were settled, matters would be such that he could put all the affairs in Henry Rymill's care and make that trip to Europe 'which has been postponed so many times'. He could expect to see Graham and his family in about 18 months, if he were spared.

Health doubts intruded upon this optimistic scenario as early as the following month, when he described himself as having been an invalid and, as Henry

Rymill reinforced to his brother-in-law, it added urgency to Ayers' planning for his trip. Apologizing for not having written recently, young Henry gives as the main reason an increase in workload in the SAMA office due to the reduction in staff; cost-cutting Ayers had let the junior clerk go. Rymill also cited the increased load of dealing with Ayers' private matters. 'Between ourselves', he wrote to Graham, matters had to be put in order to allow the visit to go ahead with a departure from the colony planned for January 1864 – about 14 months hence:

> I think he now begins to find that of the delays going much longer his constitution may become so . . . [illegible] that even if he were so inclined by and by to take a run more than half the pleasure of it may be done away with in consequence of indisposition.

Another problem was still to be dealt with. Walters had written on behalf of the Henderson people and it looked less likely that an agreement could be made. Henry Rymill's comment was: 'the Company at home are so grasping and dictatorial that I think little or nothing can come of it with them at the helm'. A possible alternative Ayers saw was to trial another process for dealing with poor-quality ores, developed by the local Captain Richard Rodda.

Representing the Colony

On the political front, Ayers was under pressure to become 'Prime Minister'. However, he did not feel the time was right. The terminology Ayers uses to Graham is interesting; Prime Minister was not the term normally employed in the public arena for the head of South Australian governments. The Chief Secretary on occasion was recognized as leader of government, though, as P.A. Howell (1986) has discussed, the leading minister was not always the Chief Secretary, and vice versa. As Ayers himself said, no matter who had been called by the Governor to form a ministry, Cabinet members would themselves agree leadership. As late as the middle of 1870 and beyond, the then Chief Secretary was questioned in the Legislative Council as to whether he led the government. Was he the minister in closest touch with the Governor and, if he resigned, would the government fall as well? It could be a complex question, and the term 'Premier', which more recently has denoted the leader of government in the Australian colonies/states, was only occasionally used in Ayers' day by commentators. A future Governor's attempt to award the title officially would precipitate a general election. Given the fact that Ayers had had no ministerial experience in anyone else's government, by

1862, it was probably a very wise move to hold off from taking the top office.

Nevertheless, early in 1863 Ayers found himself virtually in the front rank, as one of a three-member South Australian delegation to an intercolonial conference to be held in Sydney to discuss, among other issues, unifying the colonies' customs tariffs. Doubtless, Ayers' business acumen singled him out as a strong representative of his colony's interests. However, the conference was called off because Victoria could not agree on time or venue. In efforts by Chief Secretary G.M. Waterhouse to rearrange the conference, the 'Tyranny of Distance' – in Geoffrey Blainey's (1993) famous phrase – was somewhat mitigated by the use of telegrams for at least some of the communications. By that period, some 4,000 miles of telegraph wire linked the eastern colonies. Through Waterhouse's patience, and to his relief, a date was fixed, with the conference to be held from late March into the April of 1863, in Melbourne.

A newspaper cutting kept by Ayers reveals a journalist's raised eyebrow at South Australia's not actually sending a ministerial team. This was not strictly true as one of the representatives was Treasurer Arthur Blyth. If the newspaper considered the South Australian representatives to have been something of a 'B' Team, at least the reporter was gracious enough to describe them as 'competent'. Ayers was given a brief biographical sketch for the sake of the other colonies' readers who would not have known him. The Burra's success was credited largely to Ayers' financial handling. This commercial success and his membership of the upper house were described as his only spheres of involvement – although by then he was also on the boards of the Savings Bank, *The South Australian Advertiser*, the Bank of Australasia, and the South Australian Gas Company. As far as the political sphere was concerned, it was known that he had already turned down office, and was believed to be biding his time, as the journalist described it, being in any case 'on the right side of fifty'. In parliament, he was recognized as a cogent arguer, if an infrequent debater. His political leaning was towards the conservative side which, 'considering the level to which he [had] raised himself, [was] not to be wondered at'. However, having said that, he was 'very liberal with his own purse as well as his sentiment'. Where the reporter had got this impression from – liberality with his purse – is not clear. Certainly, Ayers has left behind something of a reputation for holding his own purse-strings pretty tightly.

The third member of the colony's delegation rejoiced in a name which would have admirably fitted him as a character in any Dickens novel: Lavington Glyde. Ayers' name was noted as having moved a resumption of discussion on the possibility of setting up an intercolonial Court of Appeal. This very much accords with Ayers' interests; he had been on the committee drawing up a reply to the Governor's Speech opening the 1861 parliament, which included a suggestion for a Court of Appeal within the colony. Certainly, to this point, legal and financial questions emerge from the parliamentary records as the main centres of interest and involvement for Ayers – he was playing to his strengths – and he had again called for papers to be tabled on the issue of a possible Australian Appeals Court, a couple of months before the conference was mooted.

Suffice it to say at this point that Ayers' attempt to keep free of high office would soon prove unsuccessful: in July 1863 he was persuaded to become Chief Secretary, or Premier. Forestalling an adverse reaction from Graham, he said he expected to be out as quickly as he went in, in all probability before his letter would leave! Graham indeed 'misapproved', concerned that a pre-occupied Henry Ayers might pay less close attention to the magnate's business. Ayers sought to reassure by saying he would not let the irons (of business) go cold. Graham's disappointment was clearly on his own behalf, and not at all for Ayers' sake, that the longed-for trip Home which 'Nothing but business or death' would prevent, was further postponed.

At this time, Ayers was instrumental in enlightening the colony literally, as well as, hopefully, figuratively. The busiest city precincts were lit by gas for the first time, due in no small part to the promotion of a South Australian Gas Company by Ayers and others. As Chairman of the Board, he had laid the foundation stone of the gas works at Brompton the previous Christmas – what a delight to the eye, and nose, that works would have been to whoever occupied Prospect House!

Ayers had also moved the readings of a bill to provide a fire brigade in Adelaide, which had passed in the middle of 1862, and a Bill for Water Works. If fire and water had been dealt with, there remained a panacea to cope with South Australia's ferocious summer heat. Ayers moved readings of a bill to give a patent to a James Harrison for his ice-making process. All told, not a bad Session for public improvements from a backbencher.

An Air of Nostalgia

On the business front, late 1862 and the first half of 1863 appeared quite satisfactory. The Fisher rescue was settled and Ayers struck a new deal with the trusty E&ACC. He rationalized disappointment, not only with the Henderson people and their process, but also with Captain Rodda's local attempt at innovation, declaring to Graham: 'I begin to lose all faith in these experimentalists and believe there is no system yet discerned which will supersede Smelting by coal.'

A cursory remark by Ayers in his May letter to Graham, in which he simply stated that Duff was unable to pay anything he owed to the magnate, so that

Ayers had repossessed his property, covered quite a touching personal story. It concerned one of the earliest Pioneer colonists – a breed Ayers would come to defend and support, in their hard times. A parliamentary petition in 1861 had called for public support for a man who had, on his own responsibility, virtually saved the earliest arrivals and Light's surveying party back in 1836. Duff had captained his own ship, *Africaine*, to South Australia and, finding the pioneers almost destitute of supplies, had gone on to Hobart and procured £3,500 worth of fresh food – backed solely by his own resources, having no official authority. Now elderly and in poor health, he struggled to hold down the posts of Harbor Master and Lightkeeper at Holdfast Bay. His trading ventures had not paid off and the petitioners felt the least a grateful colony could do was to make some gesture of support from the public purse.

Though business matters were, naturally, to the fore, personal matters also figured prominently in the Ayers/Graham correspondence through the first half of 1863. Ayers thanked Graham for the photos of his children. In return, he sent his 'Carte de Visite' for Graham to see whether he had 'worn well' in the five years since Graham's visit.

An air of stocktaking and reminiscence came over him. His twenty-third wedding anniversary was approaching, he told Graham, and son Frank had just had his twenty-first birthday. Maybe a little icing was missing from the cake, as he observed that, had his eldest been a daughter, he would 'likely have been a Grandpa by this time. Verily time flies.' He gathered that Graham was now 50, and Ayers' extravagant greeting stretched to 'May you live to complete an entire hundred years.' He remarked that his own father was nearly 80. However, mortality intruded upon the circle of friends as, by the time of Ayers' July letter, he reported the death of William Younghusband in Rome, on his European tour. That must have given him serious pause, when he considered for how long he had been postponing his own trip Home, and renewed his determination to hold to his plans for a January 1864 departure.

Ayers' appetite for the trip was also whetted by a series of interior photographs of the Grahams' town residence in Frankfurt, and shots taken around their country retreat, the Schloss Handschuhsheim, near Heidelberg. These had arrived in a long-anticipated box of presents which, as expected, gave 'great satisfaction to all concerned'. Ayers reckoned the country house had to be a 'secret retreat', particularly as it appeared 'so delightfully old fashioned – I quite long to see it'. The photographs of Graham's quaint German retreat must indeed have been sending out Lorelei calls to a conscientious colonist, who was, yet, impatient to have a taste of Europe's offerings.

10

THE INS AND OUTS OF OFFICE: KEEPING ON TRACK

Ayers finally capitulated to Governor Daly's call, in mid-1863, and accepted high office. A limited commitment, initially, to become Minister without Portfolio, indeed Minister without pay, in an attempt to carry a contentious issue for F.S. Dutton's government, swiftly grew into the full Chief Secretaryship when Dutton fell. Ayers was in an invidious position, having to hold his nose in passing a customs tariff measure fundamentally at odds with what had been agreed at the Melbourne Conference. He did it, he told Graham, simply to avert a crisis and the cost of an unnecessary election. He imagined only a brief stay in office as these early ministries, before the advent of a fully fledged party system, were generally short-lived. Factions coalesced and disintegrated over this issue and that. However, Ayers would be chained to a towering in-tray for over two years. The notional leadership might change, one minute Arthur Blyth, the next Dutton again, but Ayers soldiered on, steadily leading in the upper house and recognized as 'Premier'. A Chief Secretary's £1,300 per annum was, of course, welcome while in office but the SAMA Board took the opportunity of pruning *their* Secretary's £850 to a more moderate £500.

What levers of power might this novice Chief Secretary pull? He had not, after all, warmed up with another Cabinet position, those ministries normally the preserve of lower house members. Hand-to-hand fighting would be necessary with the awkward Justice Boothby, Second Judge in the Supreme Court, who had steadily refused to accept the legality of any acts of the colonial parliament. Boothby's less hard-line fellow justices maintained that if an act had received Royal Assent, it should be accepted. Some politicians felt that laws passed by the democratically elected government should not fall prey to an unelected judge's 'interpretation', let alone his refusal to even acknowledge their authenticity. Westminster stepped in with a helpful Colonial Laws Validity Act. 'Square peg' Boothby, however, refused to accept the validity of the Validity Act and so the issue remained a headache for future solution.

The Northern Territory

A major undertaking for Ayers' first ministry was establishing a colony's colony. South Australia was allowed by the British Government to expand into its 'Northern Territory' as long as it promised to construct its section of a telegraph line to unite the colonies with the Mother Country. Ayers was bold enough to make this commitment in order to forestall Queensland getting in on the act. He wanted to ensure that Adelaide was the first to receive news, not bringing up the end of the line, so it could be kept in the loop of international happenings. The fact that this colonization was not only a government project but also, for Ayers, Graham and many others a commercial one – they were purchasing tracts of land up there – was questionable practice. Ayers teased Graham with the prospect that they might 'catch some copper or gold mines with luck'. If not, well, 'seven shillings and sixpence an acre for country land and three shillings and nine pence each for Town lots will not ruin us'. The idea of conflict of interest was only just becoming an issue to be considered both at local- and colonial-government level.

Establishing the new colony would be an enormous challenge, however. The continent had only recently been crossed, from Adelaide in the south to the Timor Sea in the north, and back, by South Australia's John McDouall Stuart, in 1862. A previous attempt by Victoria's Burke and Wills had ended tragically. Founding the Northern Territory was bound, then, to be fraught with difficulty.

Ayers must have thought he had just the man for the job in Boyle Travers Finniss. Once a young army officer, he had given up his commission to support Colonel Light's surveying efforts in the first days of South Australia. He later rose to lead government, bridging the change from the old legislature into the new. The man was therefore no slouch and expectations were high for the success of his mission. With great fanfare, and a public breakfast at the Port, Finniss, as Government Resident of the projected Northern Territory colony, prepared to lead his expeditionary party for the North aboard the *Henry Ellis*.

Ayers' signature, as Chief Secretary, was appended to all the letters of instruction given to the officers of the pioneering party. In tone and detail, they bear the hallmarks of his letters of instruction to that parade of former officers of the Burra mine. One might conclude that he may well have drafted these grander-scaled instructions, for Finniss and his subordinates, himself. With some of his own money sunk into the project, there is no doubt that he would at least have carefully scrutinized the documents, even if the initial drafting had been by civil servants; several years later, he confirmed this close involvement. Ayers probably made a mistake in attempting to 'micro-manage' Finniss' expedition. This was remote control on a continental scale. That Finniss' attention to these instructions was seen more in the breach than in the observance perhaps made failure even more likely.

Where Finniss deviated most markedly from Ayers' directive was in his response to Aborigines. Ayers wanted to achieve their acquiescence in, if not their support for, the small colony in the north, and instructed Finniss to give them gifts of counterpanes and afford them all the medical help he could manage. Instead, Finniss put them violently 'in awe'. Relations with the Aborigines had gone wrong from the start when perishable food stores left unattended on the beach were scattered around by the natives, and an Aborigine was murdered in reprisal. Finniss ordered a Coronial Inquiry and pressed the crew of the government schooner *Beatrice* into jury service – against Queen's regulations, according to Commander Hutchison, who had been having some success in building relationships with the Aborigines. An eye-for-an-eye killing of one of Finniss' officers provoked an out-and-out attack on the Aborigines under Finniss' 20-year-old son, who reputedly ordered his party to 'shoot every bloody native you see'. These appalling details would not be known until later, but an exasperated Ayers knew enough to remonstrate repeatedly with Finniss over his serious mishandling of matters and his failure to keep the government fully informed.

The Pate of a Politician

Of course, with Ayers now Chief Secretary, and with such weighty matters to deal with, it is no surprise that, when the Mail steamer left Adelaide for Britain in the much-anticipated January of 1864, he was not aboard. He turned instead to some cultural activities, in the February, stepping in to replace Anthony Forster in chairing the committee planning celebrations for the Shakespeare Tercentenary. Dramatic readings were on the agenda, representative of the bard's full range. Efforts were rewarded by a good turnout despite a 'cold and stormy' night. The Governor arrived accompanied by two of the performers:

the Chief Justice and the Chief Secretary. As a warm-up, Frank Ayers took the part of Bassanio in an excerpt from *The Merchant of Venice* and 'evinced a jealous regard for the good of his friend [Antonio]'. Frank's 'main feature' for the evening, though, was to roll them in the aisles with his Bottom.

It was time for the Chief Secretary, himself, to strut the stage, come the second half, and he appeared as First Clown/Gravedigger in the Yorick scene from *Hamlet* 'with precision, emphasis, and impressiveness'. Ayers was typecast, arguing points – 'pints' says the *Register*'s reporter, mocking the Chief Secretary's Hampshire accent – of law, having the opportunity as well to sing a couple of ditties during the scene. The clowning would have had a sharp satirical edge, too, when, on his throwing up Yorick's skull, Hamlet observes at the gravedigger's nonchalance: 'This might be the pate of a politician, which this ass now o'erreaches.' 'Ay, my lord', agreed the Horatio, played by Frank Ayers, again, performing 'creditably'. It was just a shame that the Prince of Denmark's lines were rushed, robbing the name-character of sufficient gravitas, and allowing Henry Ayers to be 'undoubtedly the "star of that goodly company"'.

Frank's company was something Ayers would soon be missing. He was sending his eldest son to London to complete his law studies, while third son, Fred, was going to Cambridge. Fred was lucky to be contemplating a future at all, having been knocked unconscious in a fall from a horse, while galloping with friends along the beach. 'Perhaps an instalment of my family in the old country may prove a strong temptation to my joining them', he suggested to J.B. Graham. Late August saw an evening party at the Ayers', as much a farewell for Frank and Fred as it was a welcome to the colony for Anne Ayers' stepsister, Elizabeth Churcher, from Canada; they had not seen her for 24 years, since she witnessed their wedding. The boys set out to pursue their studies in England four days later. The decision to arrive with the Southampton Mail leaves little doubt that they were first headed for Portsmouth, to meet their surviving grandparent, William Ayers. No doubt they would have made a pilgrimage to their grandmother Ayers' graveside, off Deadman's Lane. They would have also surely wanted to see their mother's former home in Portsea's St Mary's Street.

Intercolonial Conference

Mid-December saw Ayers and Arthur Blyth sail aboard the *Rangatira* to represent South Australia's interests at a second intercolonial conference. As a souvenir, Ayers kept a copy of a newspaper report for his scrapbook. A line-drawing illustration was reproduced showing Ayers seated at the centre of deliberations. He appears about to make a note of a point from James McCulloch of Victoria's delegation, whom he regards intently. Ayers gives the impression of a man of solid build with full beard but little or no moustache, with a mass of dark hair swept rather wildly from a right-side parting. The somewhat fragile, even Romantic, looking youth of his wedding sketch has given way to the substantial middle-aged man of affairs.

December 1864, and its Melbourne visit, saw a departure from the Ayers' traditional Christmas. Ballarat was the venue for Christmas dinner, in company with Arthur Blyth. The party appears to have been completed by daughter Maggie and Daniel Daly, nephew of the South Australian Governor, Sir Dominick. The inclusion of these two young people might be seen as an attempt to encourage a match, linking the family of the son of a Naval dockyard worker with a notable establishment figure. Of course, it may all have been more innocent than that; Maggie had only just turned 16.

Risking an Early Election

For the Chief Secretary, the time for renewal of leases on many large sheep stations gave the opportunity to reap greater government revenues from the 'squatters' – a term of gentle abuse for those occupying large pastoral leases. Numbers of squatters sat in parliament. Increasing their rents would enable the reduction of import duties for the general consumer. It had to be a sure-fire vote-winner, and Ayers bowed to temptation and chanced an early election, in March 1865, regardless of the cost. South Australia therefore went to its first election that broke the intended three-year cycle. During the campaign, Ayers was cleverly satirized in the guise of Professor Ayre of Massachusetts. The Professor's 'Constitutional Pills' were a cure-all; they were that 'useful polectikal artikel – soft soap'. They produced a great country, cheap taxes, no problems with 'squatters', roads, or education. The panacea cost more in New South Wales and Victoria, where it was sure to go up further in price, and could not even be got in Western Australia. The whole thing was an inspired and lengthy political lampoon, by 'Yankie Doodle'.

A more serious threat to electoral victory had an American connection, over the colony's defence. Ayers' administration had already been criticized for keeping under wraps just how defenceless South Australia's coastline really was, like a doctor keeping from the patient how serious his condition was for fear the news would kill him. Heightening this concern, the Melbourne government had to shoo away a Confederate warship from Port Phillip Bay.

Around this time, Frank Rymill returned from a nine-month Home visit over the question of what their brother John, the competent bushman, was up to with regard to their late father's estate. There had been a big dinner party for his farewell, Ayers remarking that he was a great favourite – 'deservedly' – and, further underlining his popularity, he added that 'his return will be anxiously looked for'. The returning Frank saw Ayers with fresh eyes: '[T]he work I think is beginning to tell upon him, I notice that he is looking much older'. General responsibilities and the gruelling election campaign had taken their toll.

Ayers had made the right election call, though, and he himself cruised to the top of the upper house poll from his thirteenth position eight years before. Unfortunately, his very competent leader in the lower house since a reshuffle the previous August, Attorney-General Stow, lost his seat when he could not simultaneously please his electors and do his duty by the government in court. Ayers had to take on, as Treasurer – and as good a leader in the lower house as Stow – Thomas Reynolds. 'Tea-pot' Tommy, the publicans had dubbed him, for his temperate leanings. Compromise was in order, with Ayers willing to try working with this out-and-out Free Trader. That policy direction, though, Ayers could not share, fearing that it would see South Australia another Singapore where the vastly wealthy soared above a 'coolie' class. One of Ayers' Cabinet, former Burra Foreman of Works, Philip Santo, saw a government's duty as keeping the gap between rich and poor only so wide.

Off the Rails

One perk of high office was, with your family, to join the Governor and others, including one-time Byron muse, Lady Charlotte Bacon, for luncheon aboard the visiting British warship *Falcon*, in April 1865. The special train carrying the party to Port Adelaide set off optimistically, but on a track in a state of temporary repair. The short, light train began oscillating alarmingly and soon left the track, the loose rails corkscrewing up behind it. There was an eerie silence, everything a jumble of top hats and crinolines. Fortunately, no one suffered worse than bruising and the hardier individuals eventually proceeded to lunch. At the ensuing Inquiry, the railway manager, American C.S. Hare, feeling pent up in the witness-box, maintained that the state of the track was not his business as the government had just separated responsibility for track and trains. He also suggested that Ayers had a direct hand in the calamity by telling the driver that the Governor wished to be back by 3 p.m., asking 'Do we go express?' or 'Are we to go express?' As it happened, Ayers was in the process of trying to 'privatize' the colony's small rail network to raise funds for nation-building projects. Under such circumstances, a train derailment with the Governor aboard is bad enough, but it is even worse if partial responsibility for the crash can be laid at your door. Much debate followed over the question of whether there was any difference implied in the words 'special' and 'express' and how these related to speed. The driver said he had gone at the speed he considered safe, and it was agreed that the train had only been travelling at about 35 miles an hour when it crashed. Hare suddenly realized he might soon need friends in high places, and immediately backed away from his aspersions on the Chief Secretary. Hare, not Ayers, carried the can, and was dismissed.

Could Yankie Doodle, the author of the political lampoon on Ayers, have been Hare? The amusing transcription of an American accent was spot on. It was prior to the rail crash but the division of Hare's responsibilities was imminent, to his displeasure, and he had no love for anyone who had made his money out of the Burra. Hare had once chaired the sacked Superintendent Burr's benefit night. He might have relished a jab at Ayers, risen so high.

The Tommy and Harry Story

Ayers was forced to give a nod in the direction of the squatters by setting up a Commission investigating the revaluations of the pastoral leases by Surveyor-General Goyder. Ayers warned the commissioners against considering individual cases, merely wishing them to look at the effects of drought in the Far North of South Australia. Their report might win for the squatters at least a temporary stay against the increased rents. The trouble in which the government soon found itself, however, was an abstruse question of whether these commissioners were breaking the Constitution in accepting a position of paid employment under the Crown while sitting as MPs. Any ploy was being used by Ayers' opponents to bring him down. Facing the inevitable Select Committee inquiry, Ayers displayed his usual aplomb; he could easily see the difference between reimbursement of the commissioners for their own time as opposed to their receiving remuneration for their task in the Far North. His new young firebrand of a Commissioner for Crown Lands, H.B.T. Strangways – soon to be described as a 'Prussian sort of a minister' – got a little hot under the collar before the Committee, in marked contrast to a typically self-possessed Ayers performance.

The prospects of success for Finniss' party in the Northern Territory worsened and Ayers replaced him with 'Big' John McKinlay, who had once mounted an attempted rescue of Burke and Wills. It is doubtful that Finniss expected a returning hero's welcome but what he got was not even a quiet debriefing in the Chief Secretary's office. A public Inquiry awaited the 'unsuccessful Romulus of Escape Cliffs', as he had been dubbed in the newspaper – referring to his preferred siting for a settlement in the north. That was where he proposed to position a new capital, 'Palmerston', as Ayers had designated it in honour of a 'venerable and eminent [British] statesman'; just about everyone other than Finniss favoured Port Darwin as the site.

Frank Rymill, of all people, did the probing questioning for the government. Why is not clear; Frank was Corresponding Clerk in the Postmaster-General's department. He would join the Crown Lands department, which seems more relevant to the task, but not for a few months yet. Regardless, he conducted his questioning well, perhaps a skill acquired previously in the department of the Police Magistrate. It is not as though Ayers was in a position to have preferred him to the post as, by the time of the Inquiry, Ayers was at last out of office.

What or who had brought him down? The answer was Tommy Reynolds and the squatters. The latter won a vote against the commissioners over their reimbursement/remuneration question, while Reynolds walked out of the ministry over two issues. He could not get his way over complete Free Trade. Neither could he stomach F.S. Dutton, nominal government leader, being appointed to succeed the retiring G.S. Walters as Agent General in London. October 1865 therefore saw the end to a 'chapter' in the 'Tommy and Harry' story, as the *Register* put it.

Reforming the Civil Service

Although out of office, Ayers was occupied with two projects as Christmas approached. From the backbench, he continued to propose a scheme of reform for the Civil Service which he had been working on before losing office. Partly inspired by what Victoria had done, it was designed to stem the flow of promising young men out of the service and into private business. The service should be two-tiered, requiring differing educational attainments, and pay-scales should be fixed for each of five bands in each tier. He felt the annual setting of pay for individuals, often by politicians who knew little about what they were doing, was degrading. He also proposed a superannuation scheme, the government putting in £20,000 to kick it off, with contributions from the civil servants themselves.

A well-regulated civil service had engaged his attention the year before the creation of the parliament, as one of a three-man Commission that started with the Police Department and then worked through every nook and cranny of the South Australian service, defining this, improving that. There were reams of recommendations, forms devised for this and that; nothing was too insignificant. The process had occupied much of mid-1856. The SAMA's MacDonald embezzlement issue had doubtless informed Ayers' approach. His place upon that Commission had probably been guaranteed after his previous work on yet another one, with the managers of the colony's two banks, advising the government on the best way of funding its projected city-to-port railway.

The two banks of that era had since been joined by others, but all had their head offices elsewhere, their profits draining out of the colony. To remedy this, the other project Ayers now involved himself in was the founding of a home-grown institution, the Bank of Adelaide. Close associates 'prevailed' upon him – his description – to add the lustre of his name to those setting the right tone for the new institution, which survived into the 1970s.

High office behind him, Ayers was able to continue arranging his trip Home, set for a year hence. Meanwhile, he spent Christmas 1865 happily on North Terrace, unaware that in Plumstead, on the Thames, his sister Elizabeth could not enjoy the festive season. Their father William had just died, albeit at a good 81. When the sad news arrived a couple of months later, all Ayers' talk of bending every effort towards making that trip Home must suddenly have sounded very hollow, with both parents gone.

11

THE CUCKOO EXERCISES HIS WINGS

Alongside his parliamentary duties, Ayers continued as SAMA Secretary. As early as 1863, given the planned European trip, he had taken an apparently sensible and innocuous step. He moved his second son, Harry, from the Bank of Australasia to work alongside him in the office. This was designed to relieve some of the weight of work on Henry Rymill and, at the same time, 'to get him [Harry] acquainted with my business'. It appears that Harry's role was not to be exclusively tied up with his father's affairs, although he did not join the SAMA's payroll. This had undergone a fair paring the previous year, when Henry Rymill had been cut back to £200 a year just as Florence was born, just as he had invested in Burra shares, and as his brother-in-law's request for interest hit home. It had to be unsettling.

The younger Henry looked with grave suspicion on Harry's arrival. He wrote to J.B. Graham that Ayers was going

> to have him broken in so that he may assist me, but chiefly I think to let him gain knowledge of his Father's business with a view, I suppose of his, ultimately, carrying it on – his knowledge of business matters is very limited and in my opinion it will take a year or two to enlighten him.

The role of deputy had all along been destined for Henry Rymill. The grooming process was well advanced and the tests of holding the fort during Ayers' absences passed with flying colours. Where, now, might lie his future? He coolly and wisely simply observed, biding his time, although he must also have been mulling over the perceived threat. At Christmas, relations appeared normal, the Rymills joining the Ayers for the usual seasonal celebrations on North Terrace.

On Christmas Eve, Ayers conscientiously wrote his final letter of the year to Graham, with a quite positive Half-Year Report on the Burra, but a sour note being the sacking, with regret, of the Second Captain, Matthew Bryant. He took his place in a lengthening line of Burra mine officers too fond of their drink. It was no wonder the Burra townships were fertile grounds for the activities of the evangelists of Temperance and Total Abstinence.

Ayers had a bone to pick with the Postmaster-General in Adelaide, and there was no more authoritative person to do so as the Post Office fell within the responsibilities of the Chief Secretary's department. He enquired why it might be that Graham had been receiving some letters and newspapers minus their stamps and, presumably, having to pay for them at his end. The Postmaster believed they were being removed by personnel in Graham's German Post Office and then sold. Ayers observed: 'That alarming and absurd epidemic "stamp collecting" having extended to such a length, that the premium on stamps, from an obscure place like this is sufficient to induce people to be dishonest.'

Early in 1864, Ayers sent encouraging news to Graham: Roach had good reports of the Burra, and Graham's loans were almost all repaid, including the one to Henry Rymill. Graham had cottoned on to an optimistic turn in the market and urged going deeper in the mine as the price of copper rose. Ayers, realistic as ever, pointed out that the costs would be so high as to obliterate profits from still-productive upper workings. He reminded his eager friend, once more, not to expect their old favourite to return its former bounty. The men were now working on a higher tribute rate, but their actual receipts were 'so small indeed, that I wonder how we keep them'.

Postal problems continued, with Ayers reassuring Graham that Henry Rymill had, he was certain, thanked his brother-in-law for the Rymills' share of the box of presents shipped the year before. While on family matters, Ayers noted that Graham's younger son, Malcolm, was expressing a desire to enter the church. His observations were no-nonsense: 'It is a most exalted calling, if a man feels that he can work in it for good, but as a mere profession it should not be thought of.' One can imagine that, if one of the Ayers boys had sensed such a vocation, it would soon have been stamped on in North Terrace.

Ayers' earlier description of a report as 'fudge' that Burra shares were selling for as low as £76 had not found favour with Graham. By the time the Mails had criss-crossed the world over the matter, the share price had dropped lower than that, so Ayers had some explaining to do. Graham felt that he had not been properly informed. There were renewed possibilities for the Henderson Process, although Ayers would still not concede that it might obviate the complex and costly ore preparations he had always supported. Realistically, Ayers admitted that there was no more fat to trim and news of a further drop in the price of copper emphasized the predicament.

July 1864 proved harassing for Ayers, in and out of parliament, the copper price giving further cause for concern, with SAMA shares falling to £65 each; he had instructions from Graham to buy more if they got as low as £50. Graham also wanted a repatriation of his funds, should Ayers be unable to secure a return of 10 per cent for him in South Australia, and Ayers had to admit that Graham might therefore expect to receive a considerable sum. Ayers would have been painfully aware that his own livelihood, and the comfort of his family, would feel the pinch of poorer returns from the Burra plus a stiff reduction in his friend's funds under Ayers' management. His hopes were, for the moment, pinned on a good deal with the Henderson people.

Rymill Rumblings

Before the Ayers brothers arrived safely in Britain, a storm of sorts blew up at home. A year and a half had passed since Harry Ayers' arrival in the office. Young Henry Rymill's monitoring of what he perceived as an increasingly threatening predicament finally reached the stage where he could keep silent no longer and, in the September of 1864, he wrote to his sister and brother-in-law to report how things now lay between himself and Henry Ayers. The letter makes unhappy reading from many points of view.

Rymill reported that Ayers' Frank and Fred were expected to be away for a five-year period, Anne Ayers, at least, 'sadly cut up at parting with them'. By contrast, the composure of Mr Ayers did not seem greatly affected, 'which did not surprise me much, for he seldom or ever shows any great amount of feeling'. In this, then, Ayers appeared the classic stern Victorian husband and father, bearing a stiff upper lip in any adversity. This critical reference to a lack of emotional display by Ayers is the first slightly derogatory remark that Henry Rymill expressed openly about his mentor. Until then, everything was Mr Ayers this, and Mr Ayers that, and always complimentary and admiring. At last, though, the acolyte was starting to recognize that the idol had feet of clay. When he used the terms 'seldom or ever' regarding Ayers' emotional expression, one wonders if he recalled, at that moment of dissatisfaction, Ayers' near breakdown at the death of little Sydney Breaks Ayers. Henry Rymill had then reported himself as the one who took hold of the situation and, through the suggested trip to Melbourne for the grieving parents, had staved off a complete collapse of this 'unemotional' man.

Having once begun with a gentle tilt at Ayers, Rymill went on to the matter at the heart of his disillusionment. He wrote:

> ... as an illustration of which [Ayers' unemotionalism] I must tell you that for a long time past I have been thinking of asking him whether he really looked to me, as he has all along said he did, to carry on his business if ever he should think fit to leave South Australia and also to have (knowing he is very close) all of [the] allowance he thought of making me for doing it: so last week I mentioned the subject to him stating my reasons for doing so were that his son Harry was each day becoming more acquainted and proudly he might look ... [illegible] by Ayers to take his place, and further that I had a family now growing up and it behoved me to prepare for the future: at the same time taking the opportunity of letting him know I considered his allowance to me of late years had not been what I sought between ourselves you know both Frank and myself have all along thought him very stingy in this respect – I am only now in the receipt of £120 per annum. From him after serving him, I was going to say night and day for nine years, and then, too without ever once during that period ever giving him any cause to murmur one word of disapprobation – I told him if I continued a single ... [unclear – man?] the increase I sought from him should not be such an object to me, but now it was a different matter – to which he <u>kindly</u> replied that if I chose to marry that was nothing to do with him – at the same time he said if I should have what I wanted he would consider it. Well as you may suppose I was not altogether pleased with his ... [illegible]
>
> Either to make one's services (and that too after such as I have rendered him) a matter of fact, I left him to ... [unclear] [h]ome, this was nearly a week ago. He has never alluded to the ... [illegible] Subject (i.e. increase or future arrangements) since, or will he; he will do so at the end of this month being the Termination of my quarter – £50 or a £100 is nothing to him especially since now making 12 per cent of account ... [illegible] his money, and yet do you know that love of making money with him is so great that he cannot bear the idea of assisting even a friend with a farthing more than he can possibly do with – if he says anything further on the subject before I close this I will add it in the P.S. I forgot to mention, however that in the course of the conversation with him, he did say his wish was to take over the management of the Burra business and do away with the present staff, proposing, he says, so that I might under his direction do the management but <u>actually</u> so that in doing this for me I would do his business (or I would be expected to) for

nothing, or next to it: but this I will not attempt, if he can not afford to pay me for doing his business (which is one of itself) so then I will see what can be done in other quarters.

Henry Rymill was not the greatest draftsman of sentences at the best of times, but this shapeless torrent represents the release of the pent-up frustrations of the previous 18 months. The bitterness is very sad after so many years of seemingly happy association and, by this stage of course, Henry Rymill was a nephew by marriage of Anne Ayers as well as brother-in-law of Henry Ayers' friend and client, John Benjamin Graham. It may well have been a classic situation of the older Henry, from the start, having seen the younger Henry as a man after his own heart whom he could mould to replace him over a period of time. Both had arrived in South Australia as 19 year olds, 15 years apart. Rymill had proven keen, willing and able to take on whatever the older man had put in front of him, much as Ayers must have appeared to his early Adelaide law-office bosses. Perhaps, too, Henry Rymill had raised Ayers' expectations of him beyond the reasonable; the more he was willing to take on, the more the older man passed his way. Numbers of times, Ayers had written in his letters to Graham that, not only Henry, but Frank Rymill as well, were beavering away at the same table, getting correspondence ready to meet the deadline for the closure of the Mail. Their efforts might be rewarded by a call for tea from Anne Ayers, but that sort of recompense would not feed a family. There may also have been a touch of over-familiarity between the two Henrys – though no suggestion of contempt bred from it on Ayers' part – with the two men in each other's company in the office day in and day out, not to mention socially as well, through long friendship backed up by family relationship. A separate working life for the younger Frank Rymill, even if he could be a willing hand in an emergency, may have helped to make Frank the 'favorite' Ayers had referred to him as, more than once. Add to this Frank's comparative youth and bachelor status, and he had all the advantages for continuing to attract the Ayers family's protective instincts which his older brother may have felt he had gradually lost.

It is difficult to gauge whether Ayers had been naive in imagining that the younger Henry would not feel slighted at the introduction of an Ayers son into the establishment, or whether the father just cynically looked to the long-term future of a son who was not destined for Law, whatever the implications for his previous protégé. The familiar situation of the reliable Rymill's always being there might have bred, if not contempt, then perhaps an easy presumption that he could always be counted upon as a second fiddle to an Ayers. Despite the passage of time, Ayers perhaps always saw Henry Rymill as a fresh-faced

1 Henry Ayers sketched at 19 at the time of his wedding, in 1840.

2 Drawing of a young Anne Ayers (née Potts).

3 Daguerreotype of Henry Ayers, 1847 or '48, taken at Burra Burra. The image appears reversed as he does not have his customary right-side parting and his waistcoat is buttoned the 'wrong' way. Yet, he says Anne regarded it as a good likeness. It also appears to be the earliest authenticated image surviving taken in South Australia.

4 Anne Ayers, photographed during a visit to Melbourne, probably in early 1867.

5 The mature Henry Ayers.

6 The mature Anne Ayers.

7 Anne Ayers' brother, Frank Potts – pivotal in turning Henry into an emigrant. If the date is correct (1863), the outdoor life has taken its toll on a 48 year old.

8 Elder daughter Maggie Ayers – 'cause' of the Rock's naming – at the age of 16 in 1864.

9 The Ayers' second son, Harry, who worked closely with his father, at the head of his favourite pony at the age of 12 in 1856. Mounted is the daughter of a Captain Meyne.

a) "All fish or that comes to net."

b) "Poor exile of Erin" Campbell

c) "Throw physic to the dogs"

10 'Heads of the People' – prominent Adelaideans of the 1840s – sketches by S.T. Gill:
  a) James Hurtle Fisher (later Sir)
  b) George Strickland 'Paddy' Kingston (later Sir)
  c) William Paxton
  d) Richard Davies Hanson (later Sir, and Chief Justice)
  e) Captain Charles Harvey Bagot.

d) "Known by his deeds"

e) "The noun Legislator — nominative case"

11 The Ayers' North Terrace house, Adelaide, c. 1860. It is probable that the two figures on the drive are Maggie Ayers and her little sister, Lucy Josephine (Josey). Possibly, Anne Ayers stands in the shade of the porch.

12 The Burra Burra mine site in 1880, three years after serious operations had ceased. The mine offices, at right, overlook the open cut of the 1870s. The pumping engine chimneys stand smokeless against the southern sky.

13 Henry Ayers' friend and client, John Benjamin Graham, the surviving correspondence with whom affords us a more than 20-year glimpse into Ayers' thoughts and attitudes, 1863.

14 Ayers acolyte-become-critic Henry Rymill, *c.* 1885.

15 Political cartoon from the *Adelaide Lantern* of Sir Henry, at the height of the 1877 political crisis, as target for naval gunnery and torpedo practice at the hands of HMS *Legislative Council*. He nevertheless stands defiant.

16 William Morgan (later Sir), SAMA Board member, fellow Unitarian and arch political foe hounding Ayers finally from government during the crisis of 1877.

19 year old and failed to realize, or fully value, what experience and expertise the young man had gained under his tutelage.

When it came down to it, blood was likely to be thicker than water and Henry Rymill was wise to seek clarification of his prospects once he saw Harry picking up the skills he himself had acquired. On the subject of his marriage, though, Rymill was on very weak ground, and his master was right to point out that that had no bearing upon the question of the remuneration for his assisting in the Ayers agency work.

On whether or not he was getting adequate recompense for his work on Ayers' private matters, one must accept his word. Clearly, he felt it was a full-time job in itself. If £120 was indeed the figure Ayers was allowing him, it was at least the figure at which Ayers had happily managed Graham's South Australian interests for five years – although Ayers had been on a larger SAMA salary than was Rymill. His observation of Ayers over the nine-year period may well have convinced him that Ayers was 'tight'. He more than most had been in a position to form such a judgement; his assessment deserves more credence than, for instance, that of the *Herald* journalist who had described Ayers as liberal with his purse. However, it was that very tightness, in SAMA matters, that had served the company well and his economies were helping, under difficult current circumstances, to keep the SAMA profitable and, incidentally, Henry Rymill in a job. It was also rather churlish to be criticizing Ayers for his tightness, when the Ayers had looked after the two Rymill brothers handsomely, taking them to the bosom of the family from their arrival in South Australia. Of course, it could all have been just policy, to keep relations sweet with the Rymills' influential brother-in-law but, if it were just a charade, the Ayers had kept it up for far longer than necessary. At any rate, when it came to the outburst, it seems that Ayers was at least amenable to looking at whatever Henry Rymill wished to propose, in the way of a rise, come the end of the quarter.

In writing to Graham of his disgruntlement, Henry Rymill made allusion to 12 per cent as the kind of interest Ayers was getting on his own money. This may have served to unsettle Graham; it was the lesser figure of 10 per cent, for instance, that Ayers had told Graham that the latter's investment in the Fisher bail-out was netting him and, by this stage, even 10 per cent was beyond what Ayers could be certain of. Perhaps Henry Rymill was merely taking an extreme example of a figure that Ayers had at one time obtained in order to make his point more forcefully as regards his superior's stinginess. Adams had once upon a time written of someone getting 20 per cent on a loan.

To round off his letter, Rymill came back to the subject of Ayers and politics, commenting that 'one and all love the emoluments of office', further tarnishing

Ayers' image, this time as just another venal politician. The Chief Secretaryship brought with it that welcome £1,300 salary, and it may be worth observing that there was no income tax in South Australia.

As Ayers contemplated this outbreak of Rymill temerity, one wonders whether it tweaked a memory of that early bid to improve his own position, 15 years before. In 1849, feeling the increasing weight of his burden of work, and with *his* young family growing, he had perhaps uncharacteristically taken an enormous risk which was probably make or break. After having to cool his heels for a period while the directors considered his demands, he had been forced to accept a smaller slice of the SAMA pie than he had been after, though a larger slice of the humble variety.

Catching the same Mail, the month's letter from Ayers to Graham included no reference at all to any difficulties between himself and Henry Rymill. He reported the good news of a new lode discovered at Burra, but also further frustrations regarding the Henderson Process negotiations. The following month, perhaps an inkling of a break-up of the old financial association between Ayers and Graham comes to light: Ayers encloses Powers of Attorney for Graham to sign, enabling Henry Rymill to receive dividends on his behalf, or to appoint another to receive them – presumably his brother, Frank. For the first time, too, Ayers' seeming unshakeable belief in the Burra falters and he is prepared to see others take on the task of ensuring its survival. It must have been quite a psychological wrench to accept the possibility of relinquishing the tiller after nearly 20 years of unwavering commitment. Ayers himself makes the suggestion that a new Cornwall-based company, founded on the most up-to-the-moment mining techniques, should take over the Burra.

Despite Ayers' business-cum-pleasure jaunt to Victoria, to the intercolonial conference, he still thought to do his duty by his friend, making up Graham's closing financial statement for the year before going away, and predating it for Christmas Eve. It is in his own hand and not prepared by deputy Rymill. All duties done on Victorian soil, the Ayers family was together again on North Terrace before the New Year of 1865 was ushered in.

If drought could affect squatters, it could also affect miners, and 1865 saw draught animals dying in droves. Timber for pit-props and pumping engines – coal likewise – could not be moved. Ayers and the SAMA Board appealed to Graham to put together a consortium in London to buy the mine. Ayers was incensed, however, when the proposed new company offered him a thousand shares if he would persuade local shareholders to vote for the deal. It was the seller, his employer, he stressed, to whom he would look for reward, not a buyer. This was a bright glint of integrity which perhaps those at the London end had not thought to spark. The whole rescue actually proved ill-timed as the world copper price plummeted. The SAMA's stock on hand just mounted.

In the April, Frank Rymill returned from his nine-month Home visit. Though ostensibly sorting out problems regarding his father's will, more likely, he went to consult with J.B. Graham over the latter's desire to remove his long-standing agency from Ayers to Henry Rymill. It would take a while for the new arrangements to even come to light, let alone fruition.

A Jittery Colony

Although electoral success had put a Home visit for Ayers further out of reach, business arrangements were constantly being refined were one to happen. In July 1866, Henry Rymill shifted from the SAMA to devote his attention entirely to Ayers' private business, assisted by young Harry Ayers, growing in experience. Harry, and Henry Rymill, were now to have the running of Ayers' and his many clients' business. Graham need not be concerned, as they would still be overseen by Ayers until late in the year. In any case, the young deputies would be working under minute instructions, drawn up in duplicate rough booklets compiled by the boss himself. These bear testimony to Lucy Lockett Ayers' description of her grandfather's meticulous working methods. They also reveal the size and range of his clientele, at something over 50. Beyond the Grahams and the Paxtons, Ayers handled the affairs of the likes of F.S. Dutton and Thomas Williams, and of a number of widows, such as that of Catherine Hayes' throat specialist, Dr Kent. Family members catered for included Elizabeth Churcher and Frank Potts. Ayers managed Sir Dominick Daly's daughter, Mrs Souttar's trust. Even such diverse figures as Captain Roach and Sir James Hurtle Fisher entrusted Ayers with their matters. Ayers managed Lucy Rymill's trust, together with her father, A.J. Baker. One imagines that Henry Rymill increasingly resented this Ayers 'intrusion' into his family's affairs.

The designated time for the handover of private business to Rymill and Harry, in July 1866, turned out quite one of the worst possible. News reached Adelaide of the stock-market crash in London, following the collapse of one-time 'bankers' bankers', Overend and Gurney. Ayers told Graham that he had been warning his friends of just such a crash for some little while. What had led him to this state of prescience, as far removed from its centre as he was, is not at all clear.

Rumour of coming war in Europe sent further jitters through the colony, its sense of defencelessness only slightly assuaged by a return visit of HMS *Falcon*, this time with HMS *Curaçoa* [sic] and Commodore of the Australia Station,

Sir William Wiseman. He and his Lady honoured the Ayers' house by dining there. Amid the balls, amateur dramatics and cricket matches of this Naval visit, it was with the Commodore that Ayers sat down. During the previous *Falcon* visit it had been with the Jack-tars that he had supped, following their more plebeian game of football. Feelings of defence reassurance were valuable because, mid-year, Bismarck strengthened his arm greatly in central Europe by the defeat of rival, Austria. Graham's German properties rang to the tramp of Prussian jackboots and spurs. Ayers had previously sent him the 'olive branches' of his family's best wishes.

Although Captain Roach made one or two better finds in the old Burra, a further drop in the copper price made working them pointless. The London crash made the prospects of the mine sale more difficult but, as Ayers remarked to Graham, the two of them had seen how swiftly these things righted themselves. Distant Adelaide could not escape the ripples, however, and, in early September, Ayers found himself chairing the wind-up of his friend Phil Levi's business. The local community had great sympathy for 'poor Phil', Ayers told Graham. The late William Younghusband's firm also collapsed and Ayers warned Graham they were not likely to be immune from the fallout. Ayers himself had an exposure of £3,000 to Levi's, though he felt well secured. Turmoil was sending tremors right down the food chain. Even an institution like the Old Scholars' Association of St Peter's sought the advice of a parent like Ayers, who counselled the committee on where its £200 of funds should be invested. Luckily, Ayers was soon able to reassure Graham that the Younghusband failure had left the magnate personally unscathed and the Bank of Adelaide, too, had held firm. If it were not for the harvest, and a good grain price, 1866 would have closed on a very low note indeed for South Australia – perhaps a period in which it was as well Ayers was not Chief Secretary.

Going Separate Ways

A happier event that year was the October marriage of Harry Ayers and Ada Fisher Morphett. A daughter of the then President of the Legislative Council, she was also a granddaughter of Sir James Hurtle Fisher. Had Morphett had his way with hereditary titles, Harry might have been marrying the Honourable Ada Morphett, or even Lady Ada. As it was, plain Miss Ada walked down the aisle for the union of two South Australian 'establishment' families. The Ayers' house resounded to five 'Evening Parties' for the young couple and their friends, but Ayers was missing Frank and Fred as he freely admitted to Graham. The fact that his own friends in Britain thought highly of the two brought 'some consolation for the loss of their society, which I feel very much'.

Feelings privately committed to the page, man to man, belie Henry Rymill's scathing criticism of Ayers as parting easily with his sons two years before.

A long and unusually hot summer stretched well into the New Year of 1867; 'I am getting very tired of South Australian summers,' Ayers complained. Though he had not been able to escape aboard the Mail steamer to Britain, he was enjoying some relief aboard the *Rangatira*. Melbourne was once again the venue for an intercolonial conference, to negotiate another cross-border tariff agreement. A new ocean postal service with Britain was also cause for South Australian concern, since a proposed west-bound route for British mail, via the isthmus of Panama, would see Adelaide at the end of the queue. The conference failed to convene, however, so the Ayers turned to sightseeing; Brunel's SS *Great Britain* was in port, and the Ayers' dancing shoes came out for a ball at the Melbourne Club. Ayers and fellow-delegate Arthur Blyth were also treated by the Victorian government to a rail excursion to 'Echuka' on the Murray. The February passage home turned out 'most pleasant', in contrast to a testing voyage six years before.

After Melbourne, Ayers had to face the fact that the Burra had run at a loss over the previous 12 months. Henderson and his process were falling by the wayside and, though other, more local 'experimentalists' tried their methods, it was to no good effect. Losses of £100 a day forced on the board the decision to cease mining. Ayers assured Graham that, if he were on the spot, he would see there was no option. The Burra population languished in unemployment, its mainstay gone. Only a few hands watched over the mothballed 'Monster', under the supervision of mine Accountant Challoner. In town, the SAMA office had to go, too, with Ayers undertaking to quietly administer for the directors, at a modest fee of £240 over the next six months, providing his own premises and clerical help.

Graham took the end of dividend payments as the signal to end Ayers' agency. Ayers was a little stung, knowing his monthly letters contained far more than just Burra matters. Conceding Graham's right, he said he would nevertheless continue to write, gratis, until dividends might resume. He did not know that Graham had long intended to shift his business to Henry Rymill.

The change of circumstances, however, flushed out a second, more intense round of Rymill disgruntlement. Ayers told Graham:

> without any previous intimation to me that he was dis-satisfied, or even disappointed at the result of our business connection, which has been in force since he left the S.A.M.A.'s employ on 1st July 1866, that he desired to make fresh arrangements with me, and, after some conversation, it was agreed that he should write to me and propose his terms, which he did, and

offered to conduct my business, in connection with his Brother Frank and my son Harry, for <u>nine hundred pounds per annum</u> for himself and brother, and I might pay my son what I liked and his services were to be continued.

Until recently, Rymill had sounded quite reasonable in acknowledging to Graham that Ayers would have concerns about his son's future. He was doubtful of Harry's experience as yet, but made it clear that Harry should take on more responsibility as his expertise grew. Then he overstepped the mark as far as Ayers was concerned. Ayers reported to Graham:

> In addition, my business which Henry and Frank Rymill influenced, or obtained should be their own, exclusively, and stipulating that certain Commissions should be charged to Borrowers for money lent on behalf of myself, or my Principals, which should be divided between themselves and myself, a practice I by no means approve and one which may lead to most serious results to the lender of the money. Because the person whose money is lent pays a Commission for having his interests looked after, which may suffer very much when the person who <u>borrows</u> pays the Agent perhaps a much larger Commission. The temptation is to overlook the interests of the employer and is a practice which ought not to prevail, as a practice, although there might be transactions where the lender might not suffer by such a course.

There are echoes here of Ayers' ire at being offered a thousand shares in the new company to facilitate the sale of the SAMA, two years earlier. He continued:

> I at once declined Henry's proposition and told him if he had any apprehension that the share he had in my business would diminish, below what it had been, viz of the rate of six hundred and seventy pounds per annum I would secure him against it ever being less than six hundred pounds a year; or, when I went to England, if he preferred a fixed income and would devote himself entirely to my business I was willing to consider what that sum should be. This he would not entertain, and intimated that as I had declined his terms, his further business connection with me must cease, and it is agreed that it is to terminate on the 31st proximo.

Discussion between Ayers and Rymill indicated agreement out of reach, and Rymill could only say that all he had proposed had been in his mentor's interest. His resignation came with feelings of 'extreme regret, after the number of years we have been together, trusting that our friendly relations will continue as heretofore and that we may still do business together.' Ayers was very displeased:

> Henry was aware that I had made arrangements when I was in Melbourne which would remove the greatest difficulty in my getting away for my Holiday and I had scarcely been forty-eight hours in Adelaide, before the agreement we had made for three or four years, was desired to be broken. I would not have let the consideration of a few hundreds a year have stood in the way of my carrying out my plan but I could not submit to having a similar business to my own, conducted by those, who were paid for doing mine, in the profit of which I was not to participate. I offer no opinion on the wisdom of Henry starting in business for himself, or of Frank giving up a Government situation of three hundred and twenty pounds a year to join him. They must each take the consequence of their own acts, and I and my Son must attend to our business ourselves. The matter has given me great pain and annoyance and I am afraid I am boring you with my long story; but as Henry and Frank came to me on their first arrival, since which I have never ceased to take an interest in their affairs, I cannot look upon this severance, which causes me so much inconvenience, with other than feelings of pain.

Naturally, this turn of events engendered great disappointment in the Ayers family that they would not now be seeing Frank and Fred as they had expected. It was nearly three years since they had left for Britain.

Death Knell for Burra?

The future – or lack of one – for the old Burra began to crystallize in Ayers' mind. Graham, Charles Beck and others should buy the mine, perhaps with the E&ACC chipping in, to implement new methods. 'Machinery in all departments' was the only hope. Ayers had a new saviour in view: John Darlington, in Britain, had innovative ideas which Ayers liked the sound of. 'Science,' said the Secretary, needed to be brought to bear 'in every way, by intelligent overseers, [or] then it must be totally abandoned.' Without the latest techniques, they would be 'simply shying away money'. Ayers admitted the utmost respect for Roach, but doubted the old faithful was capable of managing any novel regime. Stern reality was staring all the players in the face: the pincer movement of a low value for the commodity and a dearth of new, rich lodes was squeezing the life out of the old concern. A world price at some £19 lower than the average during the previous 16 years was impossible to argue with. Ayers had exhausted all options for keeping on track what had been a central enterprise in the success of the experimental colony and his own fortunes as one of its pioneers. The cash-cow could be milked no more.

12

ROYAL IMPRIMATUR

Mining ceased at Burra the day before Ayers' forty-sixth birthday and, on the day after it, he led government once more. The return to Premiership was unexpected and unlooked for. He told Graham he did not delude himself that he was not facing problems, but 'many think that I am the one to confront them. I am willing to do whatever I can for South Australia . . .' Indeed, his was the only success in forming a government following fallout from a court case on the rightful ownership of the Moonta copper mine. He appeared the man of the hour once again, the colony experiencing its seventeenth ministerial crisis in 11 years of responsible – some said, irresponsible – government. The depression of the previous year would impose serious constraints on any incoming government.

On the Northern Territory question, the *Register* speculated that 'we may suppose that if the Chief Secretary is of the same opinion as Mr. H. Ayers, he will be in favour of returning the money of purchasers, as lately suggested by the holders of land-orders'. This is an appealing separation of the private and public Ayers who, it must be remembered, had personal funds tied up in Northern Territory land-orders. As for the Judge Boothby problem, London's Privy Council had patted the issue back to the colony, waving an Act of George III, under which the Ayers ministry began holding hearings. Good news in the Governor's Speech included an expansion of public works to help alleviate unemployment. The extension of the railway to Burra would doubtless go ahead although, with revenues low, it would be financed on the back of a government bonds issue. New South Wales was being accommodating about customs duties. Not everything was gloomy, therefore.

During the Reply debate, Ayers suffered a scathing attack from John Baker. Though delivering his barbs with humour, Baker laid every evil facing South Australia at Ayers' door. He accused him of grasping for revenues by the overvaluing of pastoral leases, ruining drought-ravaged pastoralists. Here, Baker had an unacknowledged conflict of interest, as a great pastoralist himself. This did not prevent him from attacking Ayers on the Northern Territory issue, on conflict-of-interest grounds, given Ayers' private concerns there. Furthermore, Baker professed to know that on the question of reimbursement if the settlement failed, Ayers was in the opposing camp to all his Cabinet colleagues in wanting his money back – the wealthy were to benefit at the expense of ordinary wage-earners. His full firepower expended, Baker insisted that he would nevertheless support Ayers and his government, 'forgetting all past errors' if he would only become 'truly penitent for all the wrongs he had done'. This raised a laugh in the chamber.

Having listened to every member's viewpoint on the Governor's Speech, Ayers rose to sum up. As was his way, he just made a few points. He characterized Baker's speech as a cut-and-paste job from past Hansard reports. The Boothby business he gave his most serious consideration. Nor was initiating proceedings without potentially drastic implications for both Governor Daly and himself, as Boothby had warned them: they could be personally and financially liable. Given that Boothby's stance had not been without supporters of some stature, Ayers tried to set minds at rest in the chamber. He reassured concerned members that almost all his Cabinet had not supported the petitions against Boothby which had been sent Home the year before. Personally, he claimed, he viewed him as innocent until proven guilty and, as for the possible financial repercussions for himself (Ayers), he could not hold the Chief Secretaryship and be worth his salt if he were cowed by such considerations. With regard to the pastoralists' hardships, he quipped that he could not call for rain from heaven, Elijah-like. As with world copper, the low price of wool due to over-production worldwide was also bearing heavily upon the pastoral sector. As for Northern Territory problems, since most of the labour in preparing the colonization attempt had fallen upon him, he had done his utmost in an effort to ensure success.

Boothby, true to form, would not recognize hearings before the Executive Council so missed all the legal arguments by barristers for and against. The government even tried to induce the obstinate judge to go of his own accord, with the offer of a pension, but he stuck to his principles. It was all to no avail, the Governor 'amoving' him from the Supreme Court bench.

The cut and thrust of parliamentary exchange was one thing but, a few days later, Baker realized he had overstepped the mark in a private, and not so private, attack on Ayers. In the chamber, he made a Personal Statement in which he denied accusing Ayers of growing rich on the backs of absentees. It is not clear that he had actually imputed fraudulent methods on Ayers' part in dealing with the investments of the likes of Graham and Paxton, but Baker obviously felt the need to swiftly back-pedal.

William Parkin rose to defend Ayers. No member in that house would command a greater respect. He had found his own team and was quite capable of holding the reins tightly, if necessary. If he would only shake off all fear, put aside all prejudices, and avoid all stubbornness, or perhaps he ought rather say 'inflexibility' of spirit, Ayers' future would be one of greatness and glory – which raised a laugh. He believed the Chief Secretary had the true welfare of the country at heart, and would not go either too fast, or too far. Striking an even more grandiloquent note, Parkin went on: if Ayers would only chalk out a broad, bold principle of action, and follow it, he would do a good work, and would 'go down to the grave with honor on his head' – 'Hear, hear', echoed around. With such fulsome backing, Ayers was now ready to embrace expansionist policies for the colony, Baker noted, in contrast with his policy stance when last in power.

Doubtless, Baker did harbour suspicion of Ayers, who had risen from next to nothing to some wealth and the pinnacle of political power. The superior attitude of the person whose wealth was founded on the land, towards the person whose wealth was derived from trade and commerce, had certainly carried over from the old country. Baker was one who believed the partly nominated legislature had served South Australia better. The new order might be representative, but was it responsible? Perhaps Baker's enthusiastic harrying of whoever was in government was born of the fact that he had held the Chief Secretaryship for a tantalizing, fleeting, fortnight under the new Constitution. Baker obviously saw his place as inquisitor, if not the steady opponent, of governments. He was certainly not, though, one of those to whom people naturally turned when there was a political crisis, unlike Henry Ayers.

New Business Arrangements

In mid-year, Henry Rymill thanked Graham for entrusting his South Australian business interests entirely to him. He did underline the fact, though, that this transition in no way reflected badly upon the way Ayers had conducted Graham's business matters over the years. Graham could feel secure, too, from the fact that Ayers, himself, was entrusting the Rymills with his business matters during his overseas trip. There had been a squabble, however, over one matter which Henry Rymill put down to Ayers' propensity for being grasping – he was not prepared to let £50 a year 'slip'. Brother Frank had something to add on the row, telling Louisa, 'after all we have done for him', Ayers was not doing much to assist them. '[T]he only explanation I can give is that I think he cannot bear to lose anything, no matter to whom.'

Without knowing of the contents of the Rymills' letters, Ayers' own told a different story. He gathered that he was to pay any of Graham's interest and maturing principals to the Rymills, and not to reinvest them as formerly. He expressed the hope, to Graham, that the new arrangements would not alter

> in any way, I hope, the friendly relations which for so long a period subsisted between us. I am very sorry, for Henry's sake, that he imposed terms which I could not assent to, as I am perfectly satisfied his prospects with me would have led to better results than the course he has taken. I wish him well and do not hesitate to put business in his way when I can do so. I can fully appreciate your desire to serve Mrs. Graham's Brothers, and the determination you have arrived at is fair to all parties. I should not have been pleased if you had withdrawn the whole of your business from my hands, simply because Henry determined to start on his own account after the many years we had been connected together, nor would it have been prudent on your part, to have disturbed the securities which I had obtained for you. Under all circumstances I do not see that you could have arrived at a more just or equitable arrangement, and it is perfectly satisfactory to me.

He was still not aware, of course, that this gradual withdrawal of Graham's business had been planned during Frank Rymill's Home visit.

Putting the Colony's Best Foot Forward

Two factors began to lighten the burden on Ayers' ministry. In late July, the agricultural season looked promising, lifting local commercial sentiment. Secondly, adding an unexpected sheen to this renewed optimism, the colony was shortly to be honoured: South Australia was expecting a 'Royal Duke [Prince Alfred, Duke of Edinburgh] to visit', Ayers told Graham, 'and we are preparing to give him a hearty welcome'. Ayers must not have been able to believe his luck that, because of the various problems preventing his European visit, it fell to him to head the government at this auspicious moment. Never before had such an opportunity presented itself for enhancing the relationship between vulnerable colony and powerful Homeland through the medium of the Royal Family. As Britain's focus shifted to developing threats in Europe, possibly at the expense of far-flung colonies, now was the time to hog centre stage. Ayers moved the setting up of a Select Committee to draft an Address of Welcome, deftly manoeuvring himself onto it. He felt it was fitting, presuming he would be presenting it to the Prince.

Perhaps most worrying for a colony preparing to preen itself before a royal representative was the news that, had the Prince paid much attention to the poorly displayed South Australian contribution to the Paris Exhibition – which he had attended – his expectations of the colony would have been extremely low. More concerning for South Australians was the realization that Britain's pre-eminence in the industrial sphere was eroding, with France, Germany and Belgium equalling anything the Mother Country could produce. It was an awkward time, too, to be coming under close scrutiny. Not only had Boothby been amoved but the Police Commissioner had just been controversially dismissed. Those unemployed were discontented, bearding the Chief Secretary in his comfortable North Terrace lair, in August. Perhaps having learned a lesson from confronting striking miners two decades earlier, Ayers treated them courteously. He reminded them, though, that he had recently taken a stroll down to the couple of gangs set some useful tasks by the Commissioner of Public Works, on the banks of the Torrens. On enquiring after obvious absentees, he had been told they had gone to wave off the 14th [Regiment], from which he concluded their situations could not be too desperate. Both newspapers kept a diplomatic silence, in their monthly summaries for Britain, about a demonstration by 200 unemployed in Victoria Square. For the sake of appearances at the very least, the colony would have to put its best foot forward.

The loyal Address was taking some drafting, and was not presented to the chamber until early September. This was cutting it fine, considering Ayers had expected the Prince as early as late August. The agreed draft contained a number of forelock tugs but also exhibited some of the pride the colonists felt in their achievement, with an eye to sending a message back home about 'the value and importance of the Australian Colonies to the British Empire'. Either arms had been twisted, or Ayers was displaying remarkable selflessness, because he announced that, as President of the Legislative Council, John Morphett would present the loyal Address.

Was the colony's loyalty out of step with sentiment at Home? Republicanism there had certainly grown during the flight of the 'Widow of Windsor' from public life following the death of Prince Albert. Even Victoria's own Secretary, General Grey, had dubbed her the 'royal malingerer'. Those South Australian locals who of recent years had begun to share similar anti-royal sentiments, scoffing at 'old fashioned' loyalist feelings, could be sniffily characterized by the *Illustrated Adelaide Post* as having adopted a 'highly intellectual' attitude. Still, if the colonial government acknowledged this nadir in the Queen's support at Home, it did not appear to care, needing to stress its links with that Home as fervently as possible.

The Weight of Work

Even amid preparations for the forthcoming excitement, the poor old Burra still plagued Ayers, with a potential sale of the mine proving difficult when one Adelaide shareholder was shouting the figure about. The continuing surface work – sorting and upgrading the quality of the ores at grass – would in fact be paying its way, Ayers commented encouragingly, if it were only possible to actually sell copper. Shares were down to £26 and, at such a price, Ayers had bought a few. At that price, he chortled, he would not mind a few more! Ayers told Graham that 'In consequence of the low cost of the necessities of life an attempt is being made to reduce wages, the working classes are however making a great resistance thereto.'

Henry Rymill had a pen-portrait of a flagging Henry Ayers for his brother-in-law: 'he told me the other day when I met him that he found the government duties too fagging and competed with his other work and he should not think of doing it again when the current term runs out'. It had already been acknowledged that the weight of responsibilities on any Chief Secretary's shoulders was too much for one man, who had to be in the chamber in sickness and in health. One who had tried it similarly recognized it as entailing a 'large amount of fagging laborious work'. The creation of an additional Cabinet position to share the burden was mooted, but Ayers was concerned about cost. The *Register* observed that he was too much of a tactician to risk his government's majority through an expensive reform in tough times.

Henry Rymill's comment on a worn-down Ayers was made in the context of revived interest in making the overseas trip. There was a Plan B. Fred Ayers, with another two years left of his degree-course, would soldier on in Cambridge but elder brother, Frank, would be coming home. Henry Rymill continued:

> the old Gentleman thinks, I assume, he will be able to entrust the business to the Sons [Frank and Harry] but for my part, although Frank may be a careful steady fellow I should be very sorry to risk it with two such inexperienced persons.

How Harry did his job was also cause for comment, Rymill describing him as having an unfortunate manner which everyone mentioned. Here, he indulges in a bit of Home-produced snobbery and prejudice: it is 'what you would expect in one who has seen very little of the world', a lack of tact, common among 'colonial children the kind of thing is too often to be met with and what I much regret to see' – Tut! Tut! young Harry!

At least Harry could be seen to be following his father's example in taking on some public duties, doing his bit on both the Music and Programmes and the Decorations Committees for the royal visit. His father would want results from the latter when attacked in parliament for the drab appearance of banners, drooping as yet unswagged, along King William Street. Henry Rymill had no good word to say about the entire royal visit. Expenditure was excessive; it was 'easy to be liberal with other people's money'. He had already hinted to Graham that a lucky Ayers had the probable chance of a knighthood: '[W]hat does it signify about a few thousand pounds of the Country's Money being spent to obtain an honour such as this?'

This judgement is a bit harsh. Was it likely that the colony would not want to put on the best show it could? Surely, the other Australian colonies, not to mention Britain's South African possessions and New Zealand, would be doing their level best – in fact, for a while, Victorians were apprehensive that South Australia would outshine them. The *Register* was on the government's side; South Australia should not disgrace itself. It would look bad to haggle over cost. It was not a competition with Sydney or Melbourne, simply a matter of expressing loyalty.

October came, with no Prince arriving. It was a busy month for Ayers anyway as, among all his governmental preparations, and day-to-day duties, he had preparations for the usual Half-Year Meeting of the SAMA. Henry Rymill disparaged that, too. It was poorly attended by merely 'friends of the Secretary – you in England must change this management'. The enterprise was 'too much in one person's hands ... the truth is his other affairs have precluded his giving the Burra business the kind of attention it really needed to make it prosper'. He believed Ayers knew this in his heart of hearts. At one moment, Ayers had been on the point of going to Britain, together with the E&ACC's manager, to see what could be jointly accomplished. The next, he had called it off. We can note, though, that the imminent royal visit could not have influenced Ayers' thinking, here, as Ayers' trip was called off well before Queen Victoria made a decision to send 'Dearest Affie' a-voyaging. Ayers would need to have been clairvoyant to have altered his plans with the Prince's visit in mind.

Rymill's suggestion that Ayers had been too busy to give much attention to the Burra mine does not bear scrutiny, either. Ayers had other lucrative sources of income, but his perpetual struggle had been to chart a course through the Burra's mounting difficulties. One only has to follow the labyrinthine twists and turns, as charted in his monthly letters to Graham, to appreciate his persistence. It would also have been utterly out of character, given the size of his own financial interest in the old concern, that he would have given the search for sustainability anything less than his best effort. His promise to Graham of keeping the 'irons hot' while in office had not been an idle one. Now, it was the world price of copper which was the entire limiter on the venture's costs. Wearing his other hat of, not just a legislator, but 'Premier', it would have been unthinkable to have a population as large as that of the Burra townships languishing in almost universal unemployment. The *Register* claimed that the colony owed a lot to Burra, and still hoped for a greater contribution towards the colony's prosperity if the mine could be rejuvenated.

The only back-pedalling Ayers could be found guilty of was related to his personal desire to get Home. A temporary divestment of his Burra responsibilities was all he was seeking; no one had given such lengthy service as he had. Henry Rymill's Iago-like insinuations to Graham about an Ayers no longer pulling his weight were most likely driving towards an end in which the apprentice could replace the sorcerer. His tricks would, he might believe, be even more spectacular than those of the old has-been – at 46 – who had known better days.

The Prince at Last

It was unclear when the Prince might arrive. Commodore Lambert's telegram from Melbourne suggested 28 October. Other news suggested the Prince would not call at Western Australia, the money expended on Perth's welcome mat being wasted. The *Register* lamented that as time was passing, the Adelaide Hills' lush spring green was becoming a 'somewhat embrowned edition'.

Protocol and Precedence were big questions. Who should stand where? Who should ride where? Why, in a colony with no established church should Bishop Short be near the head of the queue? Why should Boothby's replacement, Judge Gwynne, be well down the list behind Chief Justice Hanson? What Ayers' government had been handed was a list of Precedence dating from Henry VIII. There was no time to quibble now; save it for a future occasion. Ayers knew there were those who were unhappy but, as for himself, it mattered not at all. He did not care whether he was first or last and would be quite happy not to feature in a carriage procession at all, were it not his public duty. What colour socks should the members wear? asked William Parkin. Any colour you like, replied Ayers, so long as they can't be seen.

Still dealing with the Northern Territory problems, squeezing the Burra rail extension in before adjournment, even conciliating the worst affected squatters, Ayers held the parliament ready at a moment's notice for the Prince's arrival.

Wanderers on the Glenelg foreshore on a late October evening swore they had seen a spray of rockets go up from a ship still way off down the Gulf. A gathered crowd began to utter 'unChristian language' when, after a good while, no ship's lights appeared. The *Advertiser* had it that Ayers had received intelligence, a week in advance, that some clowns were going to try a hoax. Whether this was true, or whether it was just an attempt to make the Chief Secretary look all-knowing, is not clear.

To rendezvous with the Prince as he finally stepped onto the Glenelg jetty, a light-as-a-feather barouche in chocolate-brown with vermilion upholstery, bought for the occasion in Victoria, slipped away from Government House carrying Governor Daly and Ayers. An ailing Sir Dominick did his best to race Prince Alfred, though needing the support under each arm of Ayers and Commander Hutchison, of the *Beatrice*. The young Prince strode up the jetty at full tilt, barely giving the welcoming party time to doff their toppers. The waiting carriage took off with equal élan, though without Ayers. The Governor relegated him to several carriages back in the procession. Did Ayers care as little for a carriage procession as he had claimed? Rymill reported it that he had turned on Daly, saying he would have that £1,500 back he had loaned the Governor!

The colony went mad and, while the in-crowd dined at Government House, ordinary folk had less-than dazzling illuminations to admire. Domestic consumers had not gone as easy on the gas as requested while over-enthusiastic gaslighters had gone into action before supplies – from Ayers' SAGasCo – were at their peak. Superintendent of Telegraphs Todd's electric searchlights nevertheless stupefied onlookers.

On a 'buster' of an afternoon, the adept young Royal laid the foundation stone of a new GPO; another at the Wesleyan's Prince Alfred College would be laid before he left. Crowds crammed the new Town Hall for two balls, with partners for the Prince most judiciously selected. Scots had their moment of glory when the Prince's piper entered, at midnight on both occasions. Caution was thrown to the winds when those who could shake an ethnic leg danced a reel to their guest's war-whoops. Taking the pattern from the banquets of London Lord Mayors, the same venue hosted a men-only dinner. Ladies, in the gallery, played the part of a mute Greek Chorus. Speech-making ran rampant, with some unable to resist reference to a possible united Australia with a hereditary Viceroy, perhaps even Prince Alfred himself – the Greek throne had already been refused on his behalf.

A traditional torch-lit procession by the colony's German community – to the consternation of some 'John Bulls' – preceded songs of the Prince's ancestral homeland from a traditional male-voice choir. Ayers shepherded dinner guests into the Government House porch to savour the serenades.

The visiting party took to the Bush near Lake Alexandrina where Murray Aborigines fulfilled a pledge to dance a corroboree – or a ringbalin in their language. The younger dancers were at pains to point out that this was simply a performance, as 'cultural tourism', and that they should not be seen as in any way primitive as their parents may have been regarded. John Baker, perhaps looking for greater authenticity, shamed himself in the public view when he called for the women to dance naked. With the press and hangers-on requested to give the Prince's party a little freedom, a couple of days' kangaroo hunting followed, under a blazing sun. Ayers, taking a breather in town, nevertheless hosted a lunch for the tars again at the Town Hall.

Oppressive heat failed to daunt either the Prince, or the *Galatea*'s band, as he compared the offerings of the colony's Grand Exhibition with that of the recent Paris model. The Volunteers received new Colours, trooped in slow and quick time, and the new Athletics Club – President Henry Ayers – put on 'manly sports' but not without one or two idiots spoiling it for everybody else, splashing bystanders with mud at the water-jump. As earlier Naval officers had done, the *Galatea*'s had to face the local cricket team, and take a thrashing. With the Burra closed and failed arrangements at the Kapunda mine, the best that could be managed on a rail trip north was lunch and 'For he's a jolly good fellow' at a Kapunda pub.

With all going so well, Ayers broached the subject of an extended stay, but it could not be. Victoria beckoned. At Port Adelaide, Prince Alfred's final duty was to see 'Miss Ayers' launch the new self-righting lifeboat, the *Lady Daly*. Ayers and his colleagues had one last dinner aboard *Galatea*, softening the blow of parting. Australia's first ever royal visit, under Ayers' management committee, was an indubitable success.

13

SLINGS AND ARROWS

Normality was dull in the Prince's wake, made worse for a Chief Secretary because the promising harvest was ruined by rust and the colony's sheep flock fell prey to scab. Copper unsaleable completed a grim triumvirate. The only bright spot for Ayers was the prospect of Frank's return. Ayers missed his sons deeply: 'there is a void in the House which nothing seems to replace. . . . But [making homes of their own] is what they have been doing since the human race began and we must submit to its continuance with the best possible grace we can.'

Hardly had the dust settled on the royal visit when Sir Dominick Daly, his ageing constitution strained by his responsibilities, breathed his last. Ayers' medico-friend, William Gosse, could do nothing for him. For the first Governor to die in office, there would unquestionably be a state funeral, the only anomaly being Sir Dominick's Catholicism. His appointment had been rather controversial in that respect but he handled the situation deftly and was generally loved. Ayers' government stumped up for the black draperies enveloping the interior of St Francis Xavier's cathedral for the five-hour obsequies. Dressed in full mourning, in excruciating February heat, the Chief Secretary, as principal mourner, trudged behind the gun-carriage carrying the coffin to West Terrace cemetery's rather forlorn Catholic section. As he stood alone by the coffin in its vault, did Ayers regret his former petulance to Daly – if Rymill is to be believed – at the chocolate-brown carriage door?

Daly's death was swiftly followed by shocking news from Sydney: Prince Alfred had been shot. The would-be assassin was an Irish sympathizer, radicalized by the hangings of three Fenians outside Manchester gaol. Australians everywhere were outraged, realizing they would all be tarred by the same brush as daubed New South Welshmen. Adelaideans already knew someone was not paying attention back Home when their two defensive heavy guns had been delivered, addressed 'Adelaide, Victoria'. Ayers' legislators went swiftly into action with more loyal Addresses, carefully worded to express horror at the deed, sympathy with the Queen, and thanks to Almighty God for the failed attempt. A second went to the victim-Prince containing admirable phrases like 'indignation at the dastardly attempt'.

Northern Territory Impasse

The Northern Territory question still pressed on Ayers' government. The five-year deadline loomed for putting investors in possession of their land, and the parliament laboured over the possibility of offering reimbursements. Eventually, it was agreed that the offer should be made. However, Home investors locked up government funds lying in the London head office of the Bank of South Australia. Ayers' friend Thomas Elder had a former-colonist brother, Alexander, who, less friendly, timed his court action to cause greatest financial embarrassment for Ayers' government. A.L. Elder, as chair of a rival, London-based, North Australia Company, made sure of the biggest possible splash in *The Times*, sabotaging the sale of South Australian government bonds. F.S. Dutton, as Agent General, worked to salvage the colony's good name, saving his political protégé Ayers' bacon, having the Court release the funds.

A.L. Elder's action appears vindictive. One of the first agents in the colony that Ayers had dealt with through the SAMA, he continued to have a fair share of the ore trade until the Association selected the notably cheap James Morrison and Co. as sole agents. Perhaps Elder harboured some resentment at his company's exclusion. He had, however, contemporary cause for disgruntlement in the shape of £20,000 of his own money tied up in the Northern Territory impasse.

To speed things up in the Territory, George Goyder was seen as the man to build upon the exploratory work of his predecessors, swiftly concluding that Port Darwin was the best spot for the new north-coast outpost. He accepted, though, that, to Darwin's east, some of the land that Finniss had recommended was still worth settling. The relief in Adelaide at Goyder's apparent success was signalled by the *Illustrated Adelaide Post*: 'A few Goahead-Goyders, and a few thousand pounds, will make North Australia a veritable Java.' It also looked as though London would very soon be a mere 14 to 16 hours away from Adelaide at the end of a wire.

Even the dormant Burra stirred. A few men worked a small valuable find and Ayers sent some poor ores, sloshing about the Creek, to the Kapunda's former Captain to try a spot of acid process upon them. The Burra rail link squeaked through the Legislative Council by one vote and even a few tons of copper were actually sold, if not for much. Ayers' old Assistant, Joseph Phillips, had £150 slipped his way when he was still unemployed six months after redundancy. Ayers and the board had started to learn generosity during the 1860s: allowances had been found for two miners' widows, for example, and a bending of the rules extended support for Thomas Rogers, 'a blind man'. This was open-handed compared with the rebuff to an appeal for the widow of cave-diver Daniel Llewellyn, after the 1851 floods.

A Saviour for the Burra?

March 1868 saw Ayers 'right glad to have Frank home'. Henry Rymill gave a half-hearted welcome to the young barrister at law, finding 'a decided improvement in his general demeanour'. 'The Father and son' had called at the Rymill office, both being 'particularly agreeable'. Had Frank Ayers known it, Henry Rymill was willing to praise him to Graham for being 'so much like one of ourselves'. The ultimate accolade had been bestowed. Previously having criticized the stay-at-home colonial boy, Harry Ayers, he was now able to admire the polish attaching to Frank – in the Rymill estimation, certainly 'the best of the bunch'. The family was almost complete again and in fact growing. Harry and Ada produced the Ayers' first grandchild. Sadly, little George survived only three days, a poignant reminder for Henry and Anne of their lost Charles and Sydney.

Fighting off renewed criticism of the Adelaide directors by Graham, Ayers riposted: 'I think you judge the Directors here a little harshly, and if you should determine to visit Adelaide you will find that all has been done that could be effected for the benefit of the concern.' It was a full ten years since Graham's last visit. Would Darlington come out, or send a deputy? The Kapunda's Captain was having no success with his acid process, which would not percolate through the creek ores. The low price of copper was the worst problem of all, Ayers contended, and no one could do much about that. Long-standing SAMA Director, Thomas Waterhouse, at last determined on following his brother Home. The vacancy was filled by none other than Henry Rymill. Was this a recipe for harmony around the board table?

Ayers still strove to keep his relationship good with Graham, enquiring closely after the magnate's intentions for his children's education, though it was becoming clear that, on any overseas trip, Ayers would meet an increasingly invalid Louisa. Her brother Frank Rymill was to marry the daughter of Ayers' long-time friend 'Mr Billy' Sanders. Unusually, and considering Frank had always been a bit of a favourite, Ayers does not note this wedding in his *Memorabilia*.

In the middle of 1868, Ayers found a soul-mate when the man himself, John Darlington, arrived. He lapped up the engineer's appraisal of the Burra's problems: being a long way from anywhere, with nothing worthwhile left in the lower part of the mine. The great cost of going deeper would take years to recoup but, now Chile had 'learned to smelt', even Cornwall and Swansea were struggling. Nevertheless, Ayers was not down-in-the-mouth, heaping praise upon the departing expert, who had impressed him with his 'professional knowledge, wonderful industry and his sterling integrity'. Even at this 'eleventh hour' he might yet 'save' them. Above all, Ayers appreciated him for his sensible manner and lack of 'professional conceit'. The fact that Darlington was willing to direct innovation from London for a mere £250, until dividends could resume, endeared him even more to Ayers. Even though Darlington would be seeing Graham himself, and Ayers was up to his neck in governmental business, he summarized Darlington's lengthy report for his friend.

Ayers' Government Totters

A fall from power was again in prospect for Ayers, later in 1868. His ministry lost its support, its local land-sales policy defeated and with Northern Territory difficulties still clouding the colony's good name in London. Unable to find a successor, Officer Administering, Colonel Hamley, recalled Ayers, who obliged with an instant government. The returning Chief Secretary outlined his new programme. Ostensibly, nothing had changed; they had just grown a month older but, from his statement, it appeared that he had made a complete volte-face – to oppositionist John Baker's mind, for one. Ayers continued to face attacks over his land policy and the Northern Territory issue. Elderly Nob Captain Bagot felt a man as respected as Ayers would simply have to resign, and would probably be gone in a week. It was not quite that bad, but not much better. When, after many attacks and a defeat in the lower house, Ayers threatened an election, he was vilified for that, too, especially by John Baker.

Over the next few Sittings in both houses, the Ayers government continued under attack. Baker focused upon Ayers' abandoning his principles. Before returning to the Chief Secretaryship, Baker had understood Ayers to prefer reimbursement, with interest, for dissatisfied Northern Territory land-order holders. Now, he claimed to favour offering extra land in lieu if they would only wait a little longer. What had caused these altered policies the Council would have to judge, Baker mischievously continued. 'Was it love of office?'

Ayers might deny it. 'Was it the gain of office?' He might deny that. 'Was it love of power?' That, too, he might deny. Well, what was it then? Did he consider that he was indispensable? 'Ha! Ha!' responded Ayers. Baker thought he recollected the Chief Secretary's saying once that, when he was dead and gone, they would all estimate his loss and appreciate his services – 'No', Ayers protested. Better that they should appreciate his services there and then, Baker continued, but Ayers' services were not beneficial to the colony. Through many attacks and exchanges over what should be done about the Northern Territory, Ayers watched and listened, barely saying anything. He was prevented in any case: upper-house Standing Orders permitted the government's representative to generally speak only once during a debate.

On the local land question, Baker thought he had Ayers again. True, the new policy avoided land auctions, which encouraged high prices and favoured 'capitalists and jobbers' over purchasers of modest means. At the same time, however, Ayers did not want to make the government beholden to the farming interest. His proposal, therefore, was to allow purchase of agricultural land, 20 per cent down, and a year to pay, but Baker was convinced that this would only benefit the 'money lenders'. Invoking a New Testament feel, linking the idea of the money lenders to the tax gatherers, Baker could further attack Ayers' motives indirectly, for what was Ayers well-known as – besides the guiding hand of the Burra – if not a financier? It would be open season for the money lenders to extract 12 or 15 per cent interest from the poor struggling farmer. No, Ayers was certainly not indispensable. Again, Ayers let the attack pass without reply.

Desperately needing support in the lower house, Ayers tempted H.B.T. Strangways with a ministerial position. Strangways admitted that there was 'no member of either branch of the Legislature, nor any man in the colony, that [he] would sooner be associated with as a colleague than Mr. Ayers' – 'Hear, hear', chorused the other lower house members. However, he was 'not prepared to take office with any one who said when he thought he had the power in his own hand that he would dictate the policy he chose, and will not go a step further than that'. Quite clearly then, Ayers ran a pretty tight ship as a head of government. Strangways refused a post, believing there was a lot of 'dirty washing' which should be done at home, without his help.

A lively Session in the Legislative Council received Ayers' announcement that he had asked Colonel Hamley for a Dissolution and a fresh election. Baker's accusation now was one of venality: rather than simple resignation, Ayers was opting for caretaker mode until the election, continuing to reap a Chief Secretary's handsome reward. If Ayers had been defeated on principle, Baker continued, he could have left office with honour and respect intact.

However, he had shifted about so much on policy and so damaged himself that Baker believed he would not now be able to return to government. The *Register* consoled its readership with the belief that 'the quality of our crises is improving'. They were 'more complex, more lasting'.

Hamley thought better of holding an election and Ayers acted as caretaker pending announcement of a new ministry. Even this did not suit Baker; Ayers had not said who had been called to try to form a new ministry. That was because he knew no more than any of them could glean from the morning's papers, protested Ayers. He knew what Baker thought of him, he said, from the 'Lecture' delivered in the chamber the week before. However, out of a Council of 15 present that afternoon, Baker had not been able to get a seconder for his attack, Ayers dryly observed. William Parkin simply begged that no matter what government was in, could they just pass the Third Reading of the Northern Territory Bill and at least get matters moving there?

Still caretaking at the end of October, Ayers was surprised that J.T. Bagot could not give them any further information on progress towards a new government. Bagot disclaimed knowledge of any such developments and thought Ayers' 'hoping' with regard to a new ministry was most likely an expression of Ayers' 'hopes [lying] in an entirely different direction'. 'Laughter', notes Hansard. Bagot continued to rib the outgoing Chief Secretary by observing that he would find his hopes disappointed in this instance as in many others; laughter again. Bagot did not have to wait too much longer to fill Ayers' shoes in the Chief Secretaryship, though nominal leadership resided in the lower house in the hands of Strangways.

Even John Baker could support the new team, though he hoped Bagot had achieved the Chief Secretaryship without abandoning his principles. There had been too much of that over the previous month, he felt, and he smarted after being ridiculed in the House of Assembly for lacking a seconder in his attack on Ayers. This had been organized by the outgoing Chief Secretary, he believed.

Parkin felt a 'change of wind' was always good, even if matters were not yet settled because of the chase after office: 'If Jack was not in he was annoyed and his friends were annoyed, and if Bill was not in he and his friends were annoyed.' J.T. Bagot claimed he had abandoned no principles to get office and, when asked whether his ministry was likely to be 'permanent', he replied that of course it would not be, because he would not be ditching any principles in order to cling to power.

The first edition of the magazine *Adelaide Punch* was published shortly after these exchanges, giving the words of a duet and an accompanying full-page cartoon which depicts the two duettists. Ayers and Bagot appear like forerunners of Hinge and Bracket – albeit with beards – with Ayers as Doctor Evadne vamping at the keyboard, left hand poised in mid-air. Bagot stands in the bow of the grand pianoforte like Dame Hilda, holding aloft a spread fan with the legend 'LAND BILL' on it. The caption below is, not surprisingly,

BEWARE! 'I KNOW A MEMBER FAIR TO SEE' Duet sung with rapturous applause in the Legislative Council Chamber by Signora Bagoti and Mdlle. Aërense.

The duet ran as follows.

> BEWARE – A DUET
> As sung in the Legislative Council Chamber.
> (see Cartoon.)

B –   I know a member fair to see;
      He would in office always be,
      Foul or fair, foul or fair.
      Trust him not.
      He's schooling me! He's schooling me!

A –   He has a tongue that's soft and sound,
      Empty air! Empty air!
      His talk's a speech of 'moighty' sound.
      O beware! O beware!
      Trust him not.
      He's fooling me! He's fooling me!

B –   O he is Ayers of a golden hue,
      Burra share! Burra share!
      And what he says is nothing new.
      Take care! Take care!
      Trust him not.
      He's fooling me! He's fooling me!

A –   He talks of land schemes woven fair
      Out of air, out of air,
      Because he wanted that soft chair,
      Sitting there, sitting there.
      Trust him not
      He's full of glee! He's full of glee!
      But after Christmas we shall see.

BOTH –   O we shall see!

It is a pity that the *Adelaide Punch* only survived briefly. It might have cast amusing and pithy reflections on a period when Ayers was actually in government. As he was still topical at the time of this first edition, he also copped a gibe as a 'Heaven-born Minister'. The proof, maintained the satirist, could be found in Shakespeare: if you looked up *Hamlet*, you would find in ACT I, Scene IV, the Prince say 'Ayers, from Heaven'. This made it 'conclusive'. Perhaps, for those readers who were sufficiently erudite as to be able to complete the quote, 'blasts from hell' might lodge in their imaginations as well.

Out of Office

Chaos may have been reigning around the embattled outgoing Chief Secretary, but it does not seem to have dampened enthusiasm for the social scene, with two November evening parties. With responsibilities handed over to Bagot, he was able to consider activities impossible under the yoke of public office. He even went panning for gold, in the Barossa, with his former Attorney-General.

Henry Rymill soon told Graham of much toing and froing at the parliament, with pressure on for Ayers to form yet another government. This seemed impossible, he claimed, with Ayers set to leave public life altogether. Perhaps that was exactly how Ayers felt, just then, but there was no sign of his following through; he was not even halfway through his second term as a Member of the Legislative Council. It is true, though, that Ayers was not an enthusiastic contributor to debates through this period.

News of the death of his older brother John's wife during the October, aged 40, may have reached Ayers by year's end. Both his brothers were now widowers, William's wife having died three years earlier. The list of those departed during Ayers' absence was growing, now with two in his own generation slipping away. Thoughts very definitely turned again to the Home visit. The long-pursued mirage at last seemed possible to arrange. Thomas Elder departed for Home in April and, by then, Ayers could tell J.B. Graham of his own plan to leave in the November.

He intended resigning as SAMA Secretary, breaking a tie of all but 24 years. It proved impracticable, however, and Ayers offered the directors a short extension of his service. The problem of a suitable successor concerned him. Perhaps some Adelaide businessman might take it on – Graham may have noted no mention of Henry Rymill's name, despite his long experience of the Burra concern. The matter hung in the air. With T.G. Waterhouse gone Home and Arthur Blyth replaced by brother Neville on the SAMA Board, how big should that board be? Ayers was now of a 'modern' frame of mind; three or four 'good businessmen' would be optimal.

The Prince Returns

'We have had gay doings here recently. Another visit of H.R.H. the Duke of Edinburgh and the arrival of our new Governor', Ayers reported to Graham. Prince Alfred looked healthy and more genial than ever. The wound of the year before appeared not to have left any mark upon his constitution – in fact, his mother believed him 'decidedly improved' by the experience of an attempted assassination. Prince Alfred obviously held no hard feelings for being shot in the Antipodes; he had taken his would-be assassin's pistols home as souvenirs.

Sir James Fergusson, Bart, who was to replace the late Sir Dominick, would 'give satisfaction' in Ayers' estimation. He had received a hasty, 'take-me-as-you-find-me' sort of a welcome, said the *Register*, the same day as the Prince, once it was realized he was aboard the Mail steamer *Rangatira*. Although out of government, Ayers was invited to dine with Fergusson and Prince Alfred that evening. The party then drove to the Theatre Royal in time for the Third Act of the evening's entertainment, the play disrupted in full flight by the wild reception accorded the official party as it entered its box.

The following morning, Ayers found himself simply one of the run-of-the-mill parliamentarians present at the hand-over from Acting Governor Hamley. Fergusson was sworn in before a packed crowd in the Town Hall. A nice touch was that the new Governor and Prince Alfred drove immediately afterwards to pay their respects to Lady Daly, reduced to living in the Government Cottage at Glenelg. That evening, another dinner, at Government House, was attended by the members of the ministry, the military, the judges and 'a few other gentlemen'. There followed two levees held by the new Governor, and a ball to entertain the Prince – the only occasion upon which the ladies could 'compete on equal terms with the sterner sex . . . beyond this equality does not go', jested the newspaperman.

The Prince's earlier visit had lacked a horse race. One of the earliest amusements of South Australians was the turf. Ayers had been on the newly reformed South Australian Jockey Club committee, back in 1856. The public's interest had dropped away somewhat, though, and it was hoped that 'the Duke's Day' would boost enthusiasm for the sport of kings. A public holiday was declared and, in searing afternoon heat, the sport went ahead on the racecourse in Adelaide's East Parklands. Ayers' only note of the event was: 'very hot'.

His Royal Highness' final farewell to South Australia was as brisk and informal as his arrival. All concluded that the informality had been of his choosing, and there was a general belief that he had made this second visit purely out of friendship; there was no official call for it. The only possible disappointment left in *Galatea*'s second wake was expressed in a full-page cartoon in *Adelaide Punch*. No hoped-for knighthoods, for the President and the

Speaker, appeared forthcoming, and Morphett and Kingston were depicted as the abandoned Dido, and handmaiden, bereft on the shore as the Prince Aeneas' ship sails towards the horizon. The two gentlemen of Adelaide would have to be patient for just a little longer.

Rymill Revolt Grows

The Strangways government passed the Ayers ministry's Northern Territory Amendment Act, providing for refunds or a doubling of the acreage for those prepared to hang on to their orders. This proved popular. As a Backbencher, Ayers' main preoccupation was prison reform. For example, he favoured 'Separation', a sensible restriction of interaction between debtors, those who had committed misdemeanours, and hardened criminals. The putting down of questions on the Notice Paper on this whole subject was, itself, unusual for Ayers. He was not an inveterate inquisitor of governments – unlike a John Baker, or a Captain Bagot. He also seems to have regularly left the chamber early at this time, perhaps preparing for his impending overseas trip. Governor Fergusson provided him with introductions to both Earl Carnarvon and the Duke of Buckingham and Chandos. In the latter, we see a Governor's view of the Ayers couple:

> He is a very sensible respectable and intelligent man. who having begun with nothing has made a handsome fortune. Mrs. Ayers though not brought up as a lady is a very presentable and estimable person, and their family are well educated and agreeable members of society here.

On the family front, the Ayers finally came to lasting grandparenthood, with Harry and Ada the happy parents of Mary Elizabeth. Meanwhile, the railway was progressing towards Burra. It is indicative of how far Ayers' – and Graham's – pride and joy had fallen in its fortunes; it was not the mine's produce that was crying out for transport, but the colony's earliest staple, wool.

In June, the question of Ayers' retirement as SAMA Secretary was becoming urgent. A Frank Rymill application for the prestigious post predated an Ayers resignation, let alone the post's advertisement. Presumably, Henry Rymill, though a SAMA Director himself, judged it impossible that the post could come his way. He soon wrote to Graham of brother Frank's disappointment and frustration. Ayers again stars as the ogre.

> We both feel Ayers has acted very unfairly towards us in the matter. There is no doubt in my mind that he must have altered <u>his mind</u> just at the last

moment when he came there [to the board meeting] and any chance of Frank's securing the appointment: his reason for doing so he presented being he feared he said that we shall do better than if we had continued in his <u>employ</u>.

If a true representation of what Ayers said, it does paint him as extraordinarily petty, and Rymill's next comment is perhaps unsurprising: he had 'grave doubts Ayers is the man competent to oversee the new arrangements – in fact, I know he is not'.

If Ayers could put the knife in for the Rymills in Adelaide, Henry could retaliate by undermining the London shareholders' confidence in the outgoing Secretary. Henry Rymill implausibly maintained that Paddy Kingston had supported Ayers' position over his brother's application, returning Ayers' favour in supporting Kingston for the Speakership of the House of Assembly. If Ayers formed government again, Henry believed mining would become very secondary with him. He disparaged Ayers further for his activities of the previous couple of years, ending with the comment that Graham might have thought that Ayers would have wanted to see Frank Rymill 'distinguish himself but no it is all self with him I am afraid to say'.

Ayers did not mention to Graham any application by Frank Rymill. Perhaps Ayers discussed the possibility of a Frank Rymill Secretaryship with his old chum, Kingston, but it looks as though it went no further. The extant application does not appear to have been officially received by Ayers as Secretary. Instead, the minutes record that Ayers reconsidered his intention to resign, offering instead the option of 15 months' leave of absence, with son Harry acting meanwhile as deputy. Harry had by then clocked up six and a half years on the periphery, at least, of SAMA matters. The board acceded to this proposal. In addition, Ayers would undertake to do all he could to enhance the Association's interests while in Britain, attending the newly set up Board of Advice, and liaising with John Darlington, its designated manager. Frank Rymill had really stood no chance. Nevertheless, he left the Crown Lands office and went into partnership with his brother as agents. They were generally listed as 'accountants' in Adelaide directories.

Aggravation for Ayers came again from Graham, just before they were to reunite after an interval of 12 years. This concerned Graham's mortgage to the ailing old sea-captain, Duff, for his property at Glenelg, which Ayers was still handling. In the next Mail, Ayers had obviously received a scorcher from Graham. His reaction was to express

> deep regret and astonishment at your repeating the remark, that you were not aware I had been selling Duff's land until you received my last letter,

> when you and I have indubitable proof in our possession, that you were quite aware I had commenced selling the land <u>more than eight years ago</u>. I am the more surprised at your extra-ordinary statement, after the trouble I took to put all the particulars of the matter before you as I did in my letter of 30th June last.

Ayers went into the minutiae of the account, and the question of his commission came up again:

> I note that you 'decidedly object' to my charging any commission to Duff's Estate for all my trouble in selling the land and collecting the rents. I might have charged it nevertheless, without consulting you, and might legally do so now, for it is quite beside our arrangement. But as I never intended to charge it without your approval, I shall not debit the account, with it. At the same time, I think that the submission of my claim to your consideration should have called forth a softer phrase than the one you have used.

Each such tussle undoubtedly added its scar over the years. A disgruntled Ayers rejected an offer by Graham to fix accommodation for the Ayers family in the neighbourhood of Graham's new Hastings residence.

It must have been rather disappointing to the Ayers that Graham picked this time to put his Frankfurt property on the market, both for family reasons and as Bismarck's war clouds gathered once more over the German states and a wider Europe. St Leonard's-on-Sea was the destination for Ayers' October letter to Graham. Perhaps British sea air would do the trick for Louisa Graham's precarious health.

Headaches Before Departure

Before Ayers could leave Adelaide, the settling in of Mr Darlington's agent needed to be overseen. As had Darlington, William Swansborough impressed the directors, although Ayers was still sceptical of the benefits to be expected from a 'liquid process' up at the mine. Listlessness appears the mood of the moment – 'Little profit results from any industry.'

With Graham's business matters virtually entirely in Henry Rymill's hands, Ayers had no real need to write at Christmas. Nevertheless, he dashed off a quick note shortly before his embarkation, due in the early hours of 5 January 1870, aboard the same Mail steamer that would carry the letter. Swansborough had been unexpectedly speedy up at Burra and Ayers told Graham about a new open-cut method to be employed at the dormant mine.

A flavour of other worries attending the Ayers departure is apparent from a letter penned the same afternoon by Henry Rymill. He coldly noted for Graham that Ayers was all but gone, and that he had 'lost all regard for him'. Frank Rymill was spared any unpleasantness through being seriously ill.

Rymill's writing is difficult to read, but it seems the contentious matter concerned a mortgage to G.R. Debney, a prominent funeral undertaker. Ayers had apparently been reluctant to pass this business over to Rymill, and no wonder, as the amount concerned was substantial, at £16,000. It would have been a large commission lost. Further, Rymill accused Ayers of having cost Graham two lawyers' fees and of somehow favouring 'his friend', Tomkinson, the manager of the Bank of Australasia, by putting the business through him, again at Graham's expense. The first exchange of the day between the two Henrys achieved nothing and Ayers said he would ponder the matter and return later. Writing to Graham of Ayers' being 'mean and grasping', Rymill also reported that he had it from Ayers' clerk, James James, that discussion about the thorny issue was all that had occupied Ayers and his son, in their office, until Ayers' return to Rymill at 3.30.

The earlier meeting may have been somewhat heated, and, from Rymill's description, this was a chastened Ayers returning to the Rymill office in Imperial Chambers. The older man looked 'decidedly ill', claiming to have been suffering a 'severe attack of indigestion' since the previous day and had 'not closed his eyes all night'. Rymill believed that was 'easily accounted for'. Whatever the outcome regarding the Debney matter, Rymill was obviously anxious that Ayers might have some plausible story ready for Graham, and warned his brother-in-law to be wary, given Ayers' 'oily manner'. Whether just for the sake of form, Ayers made a last call in the Rymill office, an hour later, to say goodbye; 'nothing bad happened', Rymill added. Presumably, Ayers could not see his departure coming quickly enough.

Writing again in March 1870, we get a remarkable echo from Henry Rymill of the sentiments of his mentor, chained to a desk while others are away. How alike the two Henrys were, really; it is little wonder that they could no longer work together. Henry Rymill cannot resist a further dig at the Ayers, depicting Harry as a trier who still needed to run important documents past Rymill. To be fair, Harry did manage some business matters to the SAMA's advantage while in charge. Rymill told Graham that Harry had behaved normally with him, in his father's absence, inviting Rymill to dine at North Terrace, where he was occupying his father's house. Rymill suspected Harry's motives, certain he could be just as two-faced as his father. He wrote of 'the old man', many years before, as having spoken of 'certain persons' with 'the utmost contempt and yet the next week the persons would be dining at his House and receiving the utmost attention from him'. Rymill believed that spurning Harry's invitation would lead to the end of their 'acquaintance'. He complained that, rather, Harry '[did] not even mind in any way – this is the characteristic of the old man'. Rymill had refused on principle: not having been to the North Terrace house for two years, he would not go in Ayers' absence. He was clearly disappointed that the snub did not break up the relationship. The Ayers brothers would not actually break it off with the Rymills, concluded Henry, because they had 'too much of [their father's] tact to drop the acquaintance'.

The fact that Henry Rymill had not been near the Ayers' house for so long must have made it extremely awkward for Lucy and their children. She is not likely to have wished to cut herself and her children off from her Aunt Anne. Perhaps they resorted to meeting up elsewhere or in the men's absence.

Rymill was 'decidedly pleased' to gather that Ayers got a poor reception from Graham when they finally met up. Rymill's distilled vitriol had done its trick. He implied that Ayers had it coming for the kind of conduct of which Rymill convicted him: 'position and wealth going for very little indeed . . . in our estimation'. He was doubtless mightily relieved that Graham had not been swayed by Ayers' 'oily manner'; it all could have rebounded rather badly upon both colonial Rymills.

14

HOME, SWEET HOME?

The *Rangatira* carried the Ayers across the Bight to rendezvous with the ocean-going P&O liner *Malta* at Albany. The pilot reckoned Albany-ites indolent, all born, he commented to Ayers, in the year '<u>One thousand eight hundred and fast asleep</u>'. Ayers himself was unaccustomedly idle, lazily watching 500 tons of coal put aboard. Once the liner sailed, though, on 11 January 1870, Ayers took to his writing desk within half an hour. Only now could he believe he was headed for England.

The lengthy epistles Ayers wrote home he directs us to, in a brief entry in his *Memorabilia* for 5 January: 'Left Adelaide with Mama, Maggie, Ernest and Josey on visit to England – all details in correspondence with Frank and Harry.' These letters paint a picture of a different Ayers from the hard-working businessman and politician, as a family man, *bon viveur* and emigrant returning to cast a fresh eye over his native land.

There was some interesting company when they stopped off at Point de Galle in Ceylon (present-day Sri Lanka) to change vessels. At their hotel was a troupe of four American dwarves, led by the celebrated 'General' Tom Thumb, whom Maggie, of no great height herself, liked to sit by for meals. Ayers described their performance as an 'entertainment(?)', although he found 'Commodore' Nutt very funny both on and off stage, especially when he played pool with three long-legged officers of the 73rd Regiment, and beat them. Amid perfect sightseeing weather, Ayers admired the flowers, shrubs and trees. He saw a perfect opportunity for the South Australian Gas Company;

everywhere was lit by 'Cocoa nut oil' with even the theatre 'illuminated(?)' in the same way. The little dark-eyed children he saw reminded him of his granddaughter Mary, not yet one year old. He always mentioned her at the end of a letter, with all sorts of 'propositions for <u>borrowing</u> her' planned for their return to Adelaide.

The Ayers boarded 'a palace' of a ship, the brand-new *Hindustan*. The clientele moved up a notch, too, with an Indian prince and Sir William Hackett aboard. Time passed quickly for Ayers, with whist, though he was constantly thinking of Frank and Harry. Heat approaching the Horn of Africa provided testing conditions and Ayers wrote near the storm windows in search of fresh air. The drawback was that his foot was directly over the screw shaft, and he had to listen to a 'half-mad fellow' playing a piano who 'couldn't', while the carpenter banged away at some repair for the Mail Agent.

At the coaling station, Aden, Ayers anxiously looked for a letter from Fred. Fortunately, he did not take the postal officer's word that there was not one because, on insisting that the man turn every letter over methodically, there was the familiar handwriting. He was so delighted, he could not be bothered to remonstrate with the man. Ayers found Aden a much more important place than he had imagined, although how it sustained life he could hardly credit, with not a blade of grass. Even the stocks of coal were bleaching in the strong sun.

At Suez – the vilest place Ayers had ever been in – another letter from Fred informed his father that *The Times* had reported his being made a Companion in the Most Distinguished Order of St Michael and St George. Quite extraordinarily, Ayers made absolutely no comment on his good news. Perhaps Governor Fergusson had already dropped a hint.

The train for Alexandria chugged away on a 16-and-a-half-hour journey. Aden's heat was exchanged for bitter cold, but the Ayers were snug, fortified with oranges, biscuits and a bottle of Moselle. They saw Alexandria by carriage, but there were the same 'disgusting filth and stinks' as in Suez; he hoped that the P&O would actually be running through the Canal by the time of their return. They were relieved to board the 'sweet and clean' *Tanjore* on 13 February. A fellow passenger was none other than *The Times* journalist, made famous by his coverage of the Crimean War, W.H. Russell. It was his turn to sit next to Maggie at meals and he proved very amusing and agreeable.

Five days saw them reach Marseille where, at dawn, Fred came aboard. After five and a half years, they were glad to see him a healthy 'strapping fellow'. They took the night train to Paris, making themselves 'very jolly' despite the cold. They went to the Gaiety Theatre to see the *White Cat*. The show included three ballets, and the dancing, scenery and transformations were the finest Ayers had ever seen. Despite the weather, they took in Notre Dame and the Madeleine, but then moved on early to Boulogne. It failed to impress. It was 'not clean' and smelled strongly of fish. The family pushed on without regret, embarking at Calais for a smooth Channel crossing.

Bittersweet Reunions

The train to London took them through a Kent where the 'farms and homesteads . . . looked like pictures'. Paxton, waiting with his carriage at Victoria Station, conveyed them to the South Kensington Hotel in Queen's Gate Terrace. He was another Burra nabob who could thank Ayers for his indefatigable effort for that enterprise and his other South Australian business. Separated for 15 years, they gossiped until the early hours. They would see much of each other, though could not walk distances together because Paxton had leg-trouble. For Ayers, however, one of his earliest London purchases was a solid pair of walking boots.

An early family visit was to Ayers' sister, 'Aunt Betsey' at Plumstead, south of the Thames. This first reunion with a sibling clearly shook him. Elizabeth, nearly three years his senior, was 'much altered' though he thought he could still 'trace out something like the features [he] once knew'. All Elizabeth could detect was the sound of their brother John's voice in Henry's. This meeting, after almost 30 years a test of everyone's emotions, was only surpassed some three weeks afterwards when Ayers took the train, alone, to Portsmouth to see his eldest brother, William.

William was also much altered, ravaged by a liver and gall-bladder condition which had turned his face the colour of gold. There was no hope for him, yet his returning brother found him 'cheerful and happy, quite resigned to his condition and taking an interest in everything around him'. Doubtless, he was glad to see his brother again, but the unhappy circumstances were underlined by Ayers' great disappointment with the surroundings of his youth. He could not recall many of the places he saw, and those he could 'did not please [him]'. He could happily have died without ever having seen his old haunts again, and the same went for parts of London. Anyone sharing his railway-carriage back to the capital would probably have been struck by a brooding fellow passenger. Henry had not believed William as near his end as he proved, bronchitis taking him in July, at the age of 'fifty-seven years and five months'. Fred and his father went down for the funeral and William was laid to rest with their parents off Deadman's Lane.

Business and Pleasure

A party was held for the Ayers to renew old acquaintances. How South Australians appeared in London tended to give a rosier picture, Henry considered, of the colony's current state than was warranted. Business – personal, commercial, and political – figured prominently at this early stage, though liberally laced with dinner parties, theatre and opera-going, cultural activities, and seeing and being seen at the big events of the social calendar. Ayers was made an honorary member in such clubs as the Reform, the Conservative, the Gresham and the Athenaeum, and he took Mama and Maggie to the House of Commons – probably to see how these things were really done – for them to watch a debate from the Ladies' Gallery. It was three hours' dull stuff on the Irish Question. The Barry/Pugin rebuild of Parliament did not accord particularly well with Ayers' taste. So much repetitive small detail over a vast expanse Ayers felt detracted from the new building's required impact. To suit the legislature of a great country, Ayers looked to the authentic weight and simple grandeur of the surviving Westminster Hall. On political business, Ayers made directly for F.S. Dutton's office as Agent General, in Westminster, while Mr Darlington's City office was the venue for some SAMA business.

There, he spent a day with J.B. Graham – older-looking, he thought. Ayers was silent to Frank and Harry on any 'roasting' from Graham, if such he got, that so satisfied Henry Rymill when he heard of it. Perhaps Graham exaggerated it for Rymill's consumption, or Ayers may simply have been circumspect about mentioning such a matter in a family letter. His fabled 'tact' is indicated by his chariness about pulling his friend's leg: Graham had bought a new home, the Oxford and Cambridge Hotel, in the impressive Warrior Square, St Leonard's-on-Sea, and Ayers considered being 'wicked' in asking Graham whether he intended keeping it as a 'Pub'. Discretion dictated he 'refrain', however, although he would put money on Paxton's asking the selfsame thing. Ayers and Graham only met up whenever the SAMA's London 'Board of Advice' met. While some frostiness from Graham perhaps played a part, other problems may well have militated against social get-togethers, including Louisa's illness, the house renovations, the Grahams' absences in Germany and the need for an operation on their daughter Louisa Maude's foot.

There was also the family of Fred's fiancée, Evelyn Page, to get to know. They lived in David Garrick's old house on Royal Terrace, the Adelphi – by coincidence, the very street on which the idea of South Australia was developed. From the Pages' roof-top, Maggie and Josey watched the Prince of Wales and his sister, Princess Beatrice, officially opening the Thames Embankment in mid-July. By then, Josey was back with the family for summer holidays after studying, very happily, at a school in Bognor. Ernest, too, needed some polishing so, when amusement flagged from trailing around London with his father, the 17 year old was happy to go off for three months' English Composition in Stroud, Gloucestershire.

Earlier, father and son had visited Cohen's of Hatton Garden where both ordered 'Albert chains' made to their individual requirements and design. Cohen's also made some studs and wrist solitaires to match three carbuncle and diamond studs, presents Lady Daly had given Ayers. He spent more time in the jeweller's, and 'not a little money', with Mama and Maggie. The price of two appealing livestock paintings – £105 each – 'shut [him] up'. 'Very choice', was his tongue-in-cheek remark. He was not *that* keen. A shilling haircut with shampoo, from a 'swell' barber in a 'palace' suited his purse better.

In Mills', in Old Jewry, Ayers had his Coat of Arms and Crest finalized. Their main emblem was to be 'doves . . . so typical of the loving and peaceful disposition of the Writer', he joked to Frank and Harry. The Crest was a dove, and three appeared on the Arms: 'Enough to make a pigeon pie'. One certainly gets the feeling that he did not take this outward show too seriously.

Frank's and Harry's letters came with copies of the Adelaide papers' monthly summaries for Home consumption. Ayers began reading dutifully, although avoiding the theological reports, but soon gave up on political stuff, too. Perhaps in clearer focus at such a remove, he was confirmed in the belief that the squabbles of colonial politics were all so petty. How small they made South Australia look, and somewhat laughable. They should only report matters of public importance, he concluded, and skip tea-meetings.

On the British Social Scene

Fred's future father-in-law, among others, insisted Ayers go to Court. He protested, and in fact did not go, which perhaps indicates he was less self-aggrandizing than some supposed. Nevertheless, he celebrated the Queen's birthday as a dinner guest at Colonial Secretary Lord Granville's, wearing his Civil Service uniform, together with representatives from many colonies. Ayers was particularly taken by a very good Sauternes, and disappointed that the flunkies only allowed each guest 'a glass – and that not full. Every drop, I have no doubt is precious but as the countryman said of the cordial, "I would like some of it in a mug."'

The quality of the wines as well as the food Ayers enjoyed, at numerous luncheons and dinners, was always cause for remark to his sons, and he kept numbers of menus. On one occasion, he was principal guest at an 'epicurean' dinner at the Mercers' Company.

> [T]here is not a head ache in the Mercers' Cellar or else the libations of Hock, Sauterne, Dry and Sweet Champagne, Sherry, Burgundy and Claret, besides punch with your turtle soup, Claret cup and Loving cup, which I indulged in would not have left me in the excellent condition I feel at this moment.

One wonders how he could make the speech he did in favour of Temperance, 16 years later, as President of the South Australian Alliance. His own 'temperate' tastes nevertheless allowed several enjoyable days of wine-tasting, selecting cases for home, importers trying to take the skin off his tongue with drier and drier sherries. Unremitting slap-up dinners, despite plenty of walking, put on the pounds and, by July, even after climbing the 214 feet of the Crystal Palace tower, he weighed 13 stone nine, the heaviest he had ever been. 'I am still adding to myself', he told his sons – whether ruefully or proudly one cannot tell. He also took a wry view of himself through the eyes of young maids who came to clean and make the beds:

> I scarcely ever look up from my writing table, that I do not see several pairs of those grimy faced damsels glaring down at me, and I can fancy, from the movement of their mouths, that they are expressing some curiosity as to the kind of work 'that old buffer is a writing of'.

Early June meant the Derby. Ayers' part was to make the transport arrangements for a party of friends, while Paxton and James Fisher were seeing to the victuals. Henry very much approved their Bill of Fare; he could see there was plenty of 'sparkling' involved. From Gloucestershire, Ernest returned to join the crowd in his father's chosen wagonette pulled by four smart greys. Dr Gosse, also over from Adelaide, swelled the party, which saw Paxton's butler and Ernest overflowing onto the coachman's box of a rig whose 'turn-out', Ayers claimed, was unsurpassed all day. He had known he would have a standard to measure up to in what he could provide. The occasional brief downpour did not dampen these race-goers' spirits. Paxton and Fisher also catered excellently for universally 'vigorous appetites'. Ayers' assessment of the day was that '[i]t is the people that make the excitement, the racing is no more than other racing'. Their wagonette got caught in traffic jams on the showery way home and this boys' outing passed the time playing with pea-shooters, and 'chaffing' the occupants of other vehicles.

The family's feelings of involvement with that other British institution, the Boat Race, were of course reinforced by Fred's university link, and they were all delighted when Cambridge won, against the odds. Other outings were driving in Hyde Park's Rotten Row, with Anne and Maggie, where they would encounter other South Australians, members of the aristocracy, or even the Princess of Wales. The few quiet family evenings, when not at the theatre, or dining with friends, were spent reading and dozing, often to the gentle strummings of Maggie at the piano. Whether Ayers had much if any time for non-fiction reading under normal circumstances, on holiday he and the family lapped up Benjamin Disraeli's new release, *Lothair*, passing it on to Frank and Harry. Ayers also experienced the general shock at the death of Dickens – like himself, born of Portsmouth.

Covent Garden was a revelation, the largest and handsomest theatre Ayers had ever entered, the orchestra being a particular treat during *Lucia di Lammermoor*. It was so smooth. They were soon back for *The Magic Flute*. The 'scenery and all the auxiliaries' were very grand. The Tamino Ayers thought no better than the tenor, Squires, who sang in Lyster's company in Adelaide, though others were the best singers it had ever been Ayers' pleasure to listen to. The sweetest thing he had ever heard was diva Adelina Patti's trademark 'Home, sweet home'. Of course, he had once heard Catherine Hayes' performance. More racy was Offenbach's *Grand Duchess*, which only Fred and his father attended, although Henry found Hortense Schneider a touch mature for starring, at 37. Racier still was a half-hour's music hall grabbed by Ayers, Paxton and Louis Joseph at the Alhambra. Band and audience were 'noisy' and the latter 'naughty' besides. Ayers reeked of its smoke for 24 hours. In all his theatre going – high- or low-brow – Ayers shows himself a discerning critic.

Seats of Learning

Fred's old university beckoned the family where they were entertained by several South Australian students in their Trinity rooms, who put on a good spread for the visitors, with Ayers able to sample some 'Audit Ale'. Generally, though, he did not 'indulge in malt'. An enjoyable musical evening led Ayers to fancy that a student's life was 'not as wearisome as some persons suppose'. The visitors must have been highly satisfied to finally see the surroundings for which Fred had left them. Perhaps absorbing the atmosphere of this seat of classical learning encouraged Ayers' shift from 'Mama' to, frequently, 'Mater' when referring to Anne.

In the height of summer, Ayers took a train with Paxton and Fisher to the Royal Agricultural Show, at Oxford, joining Paxton's brother, Edmund. After they had 'refreshed [their] inner men', they examined the livestock with seasoned eyes, knowing exactly what to look for – early in Ayers' visit, he and Fisher had gone to the docks to see the Fisher brothers' wool from Bundaleer and Hill River go under the hammer. Their wool had fetched the best prices and Ayers was convinced of the competitive fairness of the auction. He and Fisher observed the knowledgeable bidding of the wool purchasers; if one of them made a bid even slightly beyond what the others considered a good price, he copped a long shrill whistle from his competitors, which made him 'look stupid'. Arriving at Edmund Paxton's farm at Willaston, Ayers was 'dead tired' and found he enjoyed his bed like never before. It was becoming a year of 'firsts'.

They took the opportunity to look round the 'seats of learning', on what Ayers felt was a 'fair Adelaide summer day' – 121°F in the sun. Ayers found Oxford lacking Cambridge's 'beautiful backs', but could have spent days in the Bodleian and Radcliffe Libraries, and cheekily took the liberty of trying out the Chancellor's chair in the Sheldonian Theatre. The party refreshed at the Mitre on cold beef and, Ayers would not say how much, sherry and ice 'with a relish only equal to what I have experienced after a long and fatiguing day at the Burra Mine'.

The family visited the Isle of Wight and took in numerous London attractions, such as Kew and the Zoo. The Ayers might go *en famille* to the Royal Academy Summer Exhibition, or to a 'Conversazione' of the Society of Arts at the South Kensington Museum. A more elaborate evening party, in fancy dress, was given by Mrs Fellows – daughter of Sir Rowland Hill, the inventor of the Penny Post and an original Colonizing Commissioner for the South Australian experiment, at whose Hampstead home the Ayers had already dined. They had a connection with Sir Rowland through his Adelaide sister, Mrs Caroline Clark. Another guest Ayers named was the widow of 'Bladen-Powell', accompanied by her two sons. Presumably one was, in fact, the then 13-year-old Robert Baden-Powell, who would go on to found the Boy Scout movement after the Boer War.

Thomas Page induced Ayers to attend meetings of the Royal Geographical Society. When the topic was Australian gold-fields, the Burra was raised and Ayers was plagued for details by a 'knot of eager listeners'. Someone's paper on his Tibetan expedition Ayers thought worthy only of a letter to the man's mother. Directly to Robert Hunt, Keeper of Mining Records at the Museum of Practical Geology, Ayers volunteered his contribution towards a government Royal Commission into the future of Britain's coal mines. From Ayers' perspective, Australia's own future coal needs would be entirely satisfied from the mines of New South Wales. Twenty-three years later, in his address to the inaugural meeting of the Australian Institute of Mining and Metallurgy – of which he was the first President – Ayers could report that he had been proven correct. That body, at its centenary, described Ayers as 'Australia's first mining industry giant'.

Great scientific names of the period hardly came any greater than that of T.H. Huxley, who had stoutly defended Darwin's *Origin of Species* against the attacks of the Bishop of Oxford – earlier the incumbent of St Mary's, Alverstoke, very shortly after the Ayers' wedding. Ayers, embracing science at every opportunity, was delighted to listen to Huxley at a Geographical Society lunch. Huxley spoke 'very nicely and has a pleasant voice. His face does not strike one, at first, as being particularly intellectual, it shows a great deal of power and his head is square and massive.' No doubt, the presence of the head of the Transatlantic Telegraph Company at the same lunch was also a strong incentive for attendance; his was a brain to be picked on the technicalities of any forthcoming telegraph link with far-distant Australia.

The Social Whirl Continues

A summer country-house weekend saw Ayers at Highclere Castle on the Berkshire-Hampshire border. He was bowled over by the grounds, especially the rhododendrons and azaleas – he was determined on growing some of the former at North Terrace. He was welcomed by Lord and Lady Carnarvon themselves and strolled with a pale and convalescing Marquess of Bristol, who sketched. With an election happening, most of the weekend guests were pleased that a Conservative had won the Isle of Wight seat, although it does not sound as though Ayers counted himself in that majority.

The British visit was notable for the absence of church-going, possibly reflecting Mr Adams' earlier sense that Ayers was not particularly focused on the after-life. On one occasion, Ayers missed church together with his friend, one-time SAMA Director John Ellis, the two men instead pinching a quiet feed of strawberries in Ellis' garden. At Lord Carnarvon's, Ayers had no choice but to attend service at the estate's church. All he could possibly look forward to, on Trinity Sunday, was an unfamiliar clergyman's sermon on that 'mystery of mysteries', Father, Son and Holy Spirit. Ayers' comparatively recent shift of allegiance to the Unitarians, at home, adds piquancy to this comment. Unexpectedly, there was no sermon. Ayers concluded: 'Perhaps his Reverence thought that there was nothing new to be advanced upon the subject of the Trinity, or that the least said thereon the better. We were thus released from church by Noon.'

In July, Ayers' hope for peace in Europe looked like being disappointed. German mines were closed and the railways commandeered, as men were called away to the colours. Wool prices plummeted, so their more recent Bundaleer product was not going to do as well as the earlier consignments. Copper was 'unsaleable'. On 16 July France declared war with Prussia.

The social round continued. Colonel Hamley called, and Ayers met Sir George Grey, the hard-retrenching young Governor who had been trying to get South Australia back on its feet when Adelaide was a very small community and Ayers a nobody. Grey would have been well aware of him, though, by the time of the 'Monster' Special Survey. Grey was waiting to discover whether he had gained a seat in the House of Commons. Disappointment would see him return to New Zealand to where he gave Ayers open invitation. A frantic dash to Richmond, before heading off to John Ellis' Kelvedon Hall, Ayers considered vital, in order to lunch with Ferdinand de Lesseps, the 'Projector' of the Suez Canal; as the two lacked a common language, Ayers chatted with the French engineer's pleasant young wife. In passing, Hyde Park appeared quite Australian. It was a most unusual season. If such good weather could be relied on *every* year, he could cope with life in Britain.

Heading North

Unexpectedly, the Ayers were to meet Thomas Williams, and his wife and daughter, not in Swansea as planned, but in Scarborough on the Yorkshire coast. The Ayers were comfortably settled, alone, into a First Class carriage after Ayers gave the guard a 'knowing look', meaning there would be a tip for him at the other end, having ensured their privacy. Heading north for York on the longest, heaviest train Ayers had ever ridden, for mile after mile there was 'scarcely a bite of grass'. He reflected that highwayman Dick Turpin would have had a very miserable time on his celebrated ride from London to York, as it was long enough, in one bout, for any human in a First Class rail carriage. The bridge over the River Ouse was one of Mr Page's, and an 'excellent construction'. Having come to prominence on the Thames Tunnel scheme under Brunel, Page had gone on to be involved with Westminster Bridge and parts of the Embankment.

The east coast seaside resort was 'simply charming'. The cliff-side gardens were filled with ladies and gentlemen, and a band played there morning and evening. The fair sex of London, in its beauty of face and dress, had already made a great impression on Ayers who described the Princess of Wales, Princess Alexandra, driving in a small pony carriage in Hyde Park, as 'the simplest and prettiest there'. In Scarborough, one young lady in particular struck Ayers quite dramatically, and he gave Frank and Harry a remarkably detailed description: she was the daughter of a Leeds manufacturer, short and not particularly good-looking. 'Her dressing surpasses anything I have ever seen', he wrote, and that is saying a great deal. The first day I saw her she appeared in Rose colored satin boots, rose-colored satin petticoat, Black silk over skirt and body trimmed with rose color satin, and black and red head dress. The next day, Rose velvet trimmed with lace, another time sea-green satin trimmed with lace. On Sunday Black silk dress, enriched with flowers, in all colors in embroidery, with a parasol that no man living can describe, and but a few women, I think.

What his sons made of this infatuation with the Yorkshire lass' presentation, one wonders. How had he reacted to a Lola Montez when 15 years younger? The turnout he so admired in Scarborough suggests a vast difference from the standard of dress he was used to seeing even on the most fashionable in South Australia.

Scarborough, they noticed, was full of German Jews and French people as it was a much safer place for them under current circumstances. He would not write more on the war, he told his sons, or the letter could be about nothing else.

The Ayers moved on to Scotland, passing many coal mines. Ayers found Newcastle-on-Tyne so 'full of smoke, fire and dirt as to make you believe it has some connection with the lower world'. Then, for the first time in their lives, they were 'in bonnie Scotland', staying in Edinburgh with an old friend and client, Robert Miller. Ayers stayed up chatting with his host until 4 a.m. – well into 12 August, the 'great slaughtering day for Grouse'. Shooting, he said, was not one of his vices, and he had already put an invitation from a military man to the back of his mind, where it remained.

Edinburgh was so rich in places of interest that the days were full of wanderings. On their first walk, the strong wind blew up dust, reminding them of 'our own beloved Adelaide'. Walter Watson Hughes was staying outside the city and together they walked around the Earl of Hopetoun's estate. This was, in its way, a link with Australia's future, because the heir soon to succeed the then Lord Hopetoun would become the first Governor-General when the Australian colonies federated.

They moved on, via Glasgow, to Loch Lomond, where the clan Paxton awaited them. Climbing Ben Lomond, Ayers was last to reach the summit – the longest four and a half miles he had ever walked – three hours up and two hours back, at the end of which he could not wait to pay his respects to the hamper, with which lame Paxton had remained, 'so soon as I came in reach of it'. The views were 'not soon to be forgotten'.

At Inveraray, and its Argyll Arms, Ayers was in for an unsuspected treat – boiled herrings. They were another something 'to be remembered' and he said he had had them for breakfast ever since he had been 'so near to them in a frizzled condition'. He confessed, he had not known what herrings were until he had tasted those of Loch Fyne.

The Argyll Arms hosted a gentleman of the cloth, whom Ayers characterized as an example of the 'muscular Christian school' because he had seen him with his jacket off, rowing his sons on Loch Lomond. This was none other than the Archbishop of York. Ayers also noticed more lowly gentlemen in holy orders touring the country and he wondered from whence they found the 'sillar' to pay for these excursions, which were 'expensive work'. They were easily recognized, with their 'wide-awake' hats, clerical shirts, single-breasted frock-coats with very little collar, and thick boots. Even out of uniform, he believed they were equally recognizable by the way they displayed impatience towards the likes of waiters, and made demands upon the good things of life.

The party moved on through stands of 'Scotch firs' up through Glen Aray. There was the odd sighting of red deer by Glen Orchy and they had a marvellous sight of Ben Cruachan. As daylight faded into a soft twilight, the Mater got somewhat alarmed, silence descending upon the whole party: they knew that one false step by a horse, or a wrong turning by their whip, would put them in a mess.

By ten o'clock, the lights of Oban finally appeared, but there was no room at the inn. The party of 30 was left standing around the luggage in the street by a tired coachman. A couple of beds were possibly available at a private house and a girl led them down a dark passageway to the foot of some winding stairs. Holyrood Palace and the murder of David Rizzio sprang to Ayers' mind, although it is unlikely that Mama wanted to hear of such a thought! They settled for a bed, a sitting room in an attic 'next the sky', and another bed in a cupboard, which Maggie 'bravely' assented to occupying, leaving the doors open for air. Ernest was put up across three chairs and a sofa at the Kings Arms. They had had nothing since three, but their landlady was good enough to set them up with ham, eggs and tea before they retired to their 'respective roosts'. Given a party of a hundred ahead of them, organized by the celebrated Thomas Cook, Ayers resolved to 'turn [his] nose southerly'.

Back in Queen's Gate Terrace, he reflected that the Scottish scenery was lovely, diverse and reminiscent of the colony, while England was more like a well-kept garden. He was pleased to find the hotel thoroughly cleaned, and with a change of management. All told, it was an improvement.

Fog Bound

Parties, theatre and shopping resumed, as the endless summer stretched on. News from France had not improved and London was full of refugees as well. Ayers witnessed the largest assemblage he had ever seen, as British supporters of the French cause demonstrated, silently, in Hyde Park. Notable fugitives were the Empress Eugénie and her son, whom the Ayers saw politely received at their Hastings hotel, while visiting Ayers' brother, John. While there, they could not resist a quick peep at Graham's Warrior Square residence, though in the magnate's absence.

An autumn country-house weekend saw Ayers with the Shropshire Leightons. The train brought him in the gathering gloom to that 'important hive of industry', Birmingham. This industrial agglomeration presented itself to him in

> the most unfavorable circumstances.... Of all the dreadful places for human creatures to pass their lives in this appears to me the worst I had seen, and I could not help thinking that I would not accept existence on the pain of living thereabouts.

Maybe, as a successful emigrant, he could survey this confronting scene and conclude that he would have had the gumption to get himself out – as had Arthur Blyth. Maybe, on the other hand, he could empathize with a local who could not see a way out, and would do away with himself. He found the 'Black Country' aptly nicknamed. As he passed mines, the sight of women landing wagons at the pit-heads and running them along tramways to the tips struck him as strange. He found a partial acceptance of these scenes: 'To one who sees this neighbourhood for the first time, his reflections cannot be very agreeable, but I suppose he would get used to it, when he knows how much the country's prosperity depends on that kind of labour.' The Leightons he found an entertaining lot but their Conservative stance, especially over the pernicious effects of charity, set him thinking.

November enveloped him in the thickest fog he had ever experienced. Landmarks a yard away were unrecognizable. People were afraid to cross the roads, cabs were running up onto the pavements, the guiding torches of 'Link Boys' were in great demand, shipping stopped on the river, and even the railways were affected. They breathed this 'pea-soup atmosphere' for days, and it was nearly as bad inside the hotel, even in bed. He told Frank and Harry he had never considered living in England but that, if he had, the previous week or two would have considerably shaken his determination.

Fred's and Evelyn Page's big day was due on the eighth and, two evenings beforehand, with Paxton back from Scotland, they all dined at the Pages'. A Mr Lazarus talked copper at length with Ayers, observing that the largest consumption of that commodity – with war continuing – was in shell cases. Ayers wrote his sons: '[W]e must all hope that the demand in that direction will soon cease.' This was hardly the sentiment of a grasping financier, and a copper-mine manager into the bargain.

Fred's 'best men' arrived for the wedding through thickening fog. It was difficult to 'see and find the church' only a couple of roads away. Even its gas-lit interior was fog-bound. Newspapers covered the event and, lest it not be reported in the Adelaide press, the proud father asked the groom's brothers to ensure a notice went in to the *Register* and the *'Tizer'* – the *Advertiser*'s nickname already recognizable in 1870.

The Nights Draw In

Departure for the family, including Fred and Evelyn, was postponed for a month, when Darlington's experiments, at Euston, were not ready. Close-by Lilywhite's could provide Ernest with cricketing gear so he walked there with his father. The weather changed further to cold, clear, crisp conditions and, by the fifteenth, light snow fell. Maybe this was sufficient to put Ayers 'out of sorts', keeping to quarters. He came face to face once more with mortality, when news came that Louisa Graham was paralyzed, her prognosis hopeless. It was the fag end of the year, the fag end of the great trip, and time for Paxton to come along for a final long business interview.

The pond in Kensington Gardens and the Serpentine provided Josey and her father with the spectacle of numerous ice-skaters. Ayers could only hope that December would not prove worse. It was. There were occasions when you could not tell the sex of passers-by at three yards, and snow lay many inches deep for weeks. Ernest likened it to treading on sugar. Ayers committed himself to remembering the 'beastly' weather when feeling 'dis-satisfied' back in Adelaide, or on hearing others complain of that city's weather. 'As it is even I am positively longing to inhale the pure air of Australia once more.'

Ayers' motto for his heraldic badge, the doves, had been selected well before this point, but his shuddering at the atmosphere of Victorian London would surely otherwise have suggested the wording: LAETO AERE FLORENT, 'they flourish in the joyful air'. Ayers had indeed flourished in the clear air of South Australia, and he now longed to escape the noisome atmosphere of a London winter.

Early illumination was remarked upon several times. This was perhaps not surprising, coming from the Chairman of the SAGasCo who, while in London, sought to discover whether the chandelier in the family dining room in North Terrace could be adapted into a gasolier. The disappointing answer was that it would then no longer match chandeliers in the other rooms. He did later have an impressive gasolier installed in his Large Dining Room. What Ayers' reaction would have been to the sight of it festooned in wet nylons, when the house was later used as a nurses' home, it is amusing to conjecture!

Through November and December there came no further signs of the looked-for peace on the Continent. Ayers baldly stated Britain's position: Russia hated 'us' because of the Crimean victories. America did likewise because of being perceived to have helped the Confederacy during the Civil War by running the Union's blockade. Prussia hated 'us' for Britain's sympathy with the French, while the French hated 'us' for not doing enough to support them. 'So we are rather more hated than beloved.'

In December, Ayers brought to fruition a labour of love arising from warm feelings for the late Dr F.C. Bayer, who had been in partnership with Dr Gosse. This well-loved Adelaide medical man had worked himself to death aged 52 in caring for high and low. Bavarian Dr Bayer had proven a prophet of his own doom; each evening after surgery, he had invariably told the Gosses: 'I am not vell.' Ayers spent time while in London helping to ensure that a bust of the late doctor would be true to the man's likeness. He was intrigued by the techniques of celebrated sculptor, and Pre-Raphaelite, Thomas Woolner, in working from a poor death cast and a small photograph – Ayers giving the artist tips, before the clay image was given some permanence in plaster of Paris, and thence into marble. It was like Dr Bayer in life, and a great comfort to his Twickenham widow, a daughter of the late Dr Kent.

On the first Jewish Sabbath in December, Ayers attended the Central Synagogue, Great Portland Street, with friend and client, Louis Joseph. The 'Chief Minister, Mr. Green', had a very fine voice, his chanting assisted by a good choir. The Jewish 'creed' was, Ayers believed, a simple one, akin to that of Unitarians: the belief in 'one Eternal unseen God'. Mr Green's 'lecture' greatly pleased Ayers, for it was 'an able one' on the 'all-engrossing topic of the day' – the people's public education, which the Jewish community supported. They had no objection to its being secular, as they saw it as very much their own religious duty to educate their children; indeed, Judaism had no higher duty. Religious education, they felt, should be taught in the home by parents, particularly the mother. Ministers (Rabbis) could only achieve so much in assisting, but it was the example lived in the home which was all-important. Ayers left, deeply impressed by the 'Reverend Green', whose discourse had been 'most liberal' and 'untrammelled by what could be called "priest-craft"'. He was an 'excellent representative of a people whose forefathers were the first to introduce intelligence into the Worship of God. And I wish the Christians of the present day would in many respects imitate their Jewish fellow citizens particularly in the case of the Poor which they manage to effect without pauperising the people.'

Josey's parents had allowed her to stay longer at Bognor when she begged them: 'All the girls say that all the life will be gone out of the school when I go because I keep them so jolly and lively.' Their bright-spark youngest was now back, loaded with presents from her school-friends who had 'deeply deplore[d] parting with her'. Christmas dinner was set at the Pages', though the best Christmas present was that the recalcitrant Paxton finally gave in and sold Ayers the North Terrace house – fulfilling a dream Harry had had months before.

Lights and Shades

Ernest and his father yielded to John Ellis, spending a few final days at Kelvedon Hall. A gossip on a cold walk through the slush with Ellis got Ayers off churchgoing again, the sun putting in an appearance – like an advertised attraction – for 'a short time only'. Ayers could not get over the difference after seeing the place in July. It was cheerless:

> The leafless trees and bushes covered in snow . . . I must have forgotten the winters of my native land or I must have seen very little of them in my youth. . . . I could not have conceived if I had not experienced it, the dreary, desolate, miserable aspect which bad winter weather produces in this country. . . . Oh, say the Farmers, we do not want much in the winter. But, what becomes of your workmen, say I, during this wretched period when they want much more food, firing, clothing. Oh, they have some of their Harvest earnings. Or, they get relief [minimal assistance] from the Parish. With all the wealth and splendour of this country, with all its industry and commercial greatness, there is an amount of pauperism and misery appalling to one, like me, who has, for the last generation, seen nothing but a genial climate and well-fed thriving people. I have at last found a punishment for all dis-satisfied Colonists. It is send them to England, so that they may arrive in the depth of winter, and let them rely on themselves for support!

One wonders if this benign picture of South Australian life was recognizable to many of the Burra's workforce. Ayers and the directors had screwed down their wages whenever possible to ensure not only the mine's survival in tough times, but also lucrative dividends, up until 1864. Even so, the miners had continued to earn above what they would have earned in Cornish mines, Ayers believed. The wages that the Burra's labourers could get were in fact double what the farm labourers in Edmund Paxton's district of Oxfordshire were getting. Ayers was really straining credulity, though, in expecting the farmers to keep their labourers on over winter while little needed doing. Perhaps Frank and Harry lifted an eyebrow at their father's indignation with the farmers, compared with his own practice – letting all the workers go when the Burra could no longer make ends meet, in 1867. Maybe the fact that the mine had recently re-opened enabled the SAMA Secretary to feel a sense of superiority over these 'tight' farmers. Or, the wealth of the Mother Country perhaps enabled Ayers to take this lofty view, in comparison with a young, relatively poor and struggling colony. At any rate, the dying year seemed to be calling more and more home-spun philosophy out of the 'old buffer'.

Before 1870 was out, Ayers rounded off his last, long letter to his sons, with a valedictory flourish to his country of birth, continuing with an apostrophe to his beloved South Australia:

> I take my leave of you now, while the fleecy snow is falling, with the earnest hope that I shall meet you well and happy in the land of your birth, and none the less my land because I was not born there, but mine from far stronger ties than the mere accident of birth. Mine, because I love it and because I believe it capable, spite of the inconveniences and drawbacks that you and I are no strangers to, of being made a prosperous and happy land. Prosperous and happy, for all its people, from its quality of climate and the fact that the necessities of life are easily procured. I shudder when I think of the misery that is existing in this country at the present moment. I am glad I have seen this country in all its phases, in its quiet beauty, in its dreary winter, in what [William] Townsend would call its 'Lights and Shades.' I am perfectly satisfied that the 'shades' more than counter-balance the 'lights,' and, for my part, I shall be content to end whatever number of days may be yet allotted to me in sunny Australia,
>
> With our united kind love to you, Ada and the babbies, and kind regards To all who ask after us.
>
>   Believe me,
>   Your affectionate Father,
>
>   Henry Ayers

15

MEANWHILE, BACK IN THE COLONY...

Shortly after Ayers left for Britain, his stand-in, Harry, managed to disregard past unpleasantness in making a party with Henry Rymill and other directors, to inspect the mine. Arriving like drowned rats and denied the comforts of a warm hotel, Harry and Rymill pressed on to the refuge of William Challoner's house. According to Rymill, Challoner's nose had been put out of joint by Darlington's lieutenant, Swansborough. Rymill saw Challoner as, generally, 'quick and conceited', maintaining he had never got on with Captain Roach. Probably worse, he saw him as an Ayers man, getting cantankerous with age. Such observations on Ayers and his associates by Rymill, writing as Graham's agent, now largely replace Ayers' own voice over the coming years: once back in the colony, Ayers either ceased writing socially to Graham or, if he did, no such letters survive.

As the rail link neared completion, the Burra townspeople wanted to make a big opening splash, with Governor Fergusson officiating. The SAMA Board, approached for a contribution, backed Chairman Kingston's proposed £25, as against Rymill's more restrained suggestion of ten guineas. Echoing his comment on Ayers' government's spending on the royal welcome, Rymill said that Kingston was liberal 'especially with other people's money'. Rymill missed the big event anyway as Lucy Rymill took ill after giving birth to a 'well performing fine little chap'.

In October 1870, for the first time since Ayers' departure, Rymill wrote a whole letter to his brother-in-law without a dig at – without a mention at all of – Ayers, or his sons. He felt neglected, though, by his own mother who he believed did 'not want to let [him] into a lot of her little secrets'. She was 'the Mother', a rather curious descriptor even for formal Victorians. Rymill was in need of solace, too, once news reached Adelaide that Louisa had lost her grip on life. The Ayers had been destined never to meet her. The circumstance of her decline conspired against any opportunity Ayers might have sought to patch things up with Graham. The magnate's absence from a farewell dinner for Ayers in the City is unsurprising given his wife's deterioration. Though we have no direct evidence, it is inconceivable that Ayers would not have sent Graham his condolences.

While away, Ayers had by no means shirked effort on the Burra's behalf. He had closely looked into Mr Darlington's acid method and, through an almost-monthly correspondence, had kept the Adelaide directors informed. Once Ayers wrote from London that he was reasonably convinced by Darlington's experiments, Rymill was amused to tell Graham how Kingston, Peacock and others on the board were suddenly in favour, too; they always left the decision making to Ayers. On-site experiments with Darlington's ore-crushing machinery were proving difficult to cost, though, giving Rymill ammunition for playing the activist at board meetings. Hearing of the Ayers' intended return, he commented that he thought the 'charms of English Society' would have induced Ayers to stay longer. The latter must have concluded 'if there [was] nothing to be made of it' he preferred 'to get back to his old quarters and screw away for the remainder of his life'. Was there a note of regret that Rymill's freedom of manoeuvre would be restricted once the 'boss' got home?

Return to Adelaide

The Ayers arrived back in Adelaide in March 1871. Had Henry Ayers been missed? Certainly not by Henry Rymill, who dryly told Graham that, due to what had happened before the visit, 'we do not know him'. The chilly Rymill welcome contrasted with that of the *Register*'s editor, who tagged a notice of the family's arrival on the end of his day's piece – which carried news of the fall of Paris and a likely end to the Franco-Prussian War. SAMA business and other matters brought the two Henrys together but, as Rymill said, he kept 'only to the particulars'. It irked him that Ayers, by contrast, showed a 'disposition to be very intimate' and was shocked at Rymill's reserve.

What would Fred Ayers do, once home? The Rymills had a 'shrewd suspicion' that the 'old gentleman' would try to get either Harry or Fred made SAMA Secretary. Rymill believed the board members 'could not say nay' to their influential Secretary, and that it would be bad for the organization.

Graham passed on his own inkling that lawyer Fred would 'turn sheep farmer'. It was not Fred who became the pastoralist but, for a wonder, father Ayers himself. In the April of 1871, Ayers bought out James Fisher's portion of Hill River Station, part of the Fisher brothers' property he helped to bail out in 1862. City-man Ayers was starting to assume at least some of the trappings of the pastoralist. Perhaps John Baker might look up to him yet!

Disproving both Rymill's and Graham's predictions, Fred joined the Adelaide law practice of Samuel Way. It was good timing, as Fred and Evelyn produced another Ayers grandchild, Frederic Gordon, to add to Harry and Ada's latest, little Henry, who greeted the returning Ayers. That other Henry, Rymill, took a swipe in two directions at once, suggesting that, after a year or so's experience, Fred might do as his brother, Frank, had done – pay £2,000 to become a partner and share in the profits.

The returning Ayers was straight back into harness as SAMA Secretary. It looks as though his clerk, James James, attended board meetings as scribe for Ayers at this stage, pre-preparing the uncontentious standard sections of the minutes. When Ayers' own careful hand takes over, outlining the board decisions, the presentation is so neat and light, even in his rough notes, that if it were not for the fact that such a thing had not been invented, one would swear he had been writing lightly in Biro.

The board passed a vote of thanks for his overseas efforts, though none for Harry's deputizing. The Secretary wanted a day set aside for discussion of future policy, when his report lasted an hour and a half. His bleakest observation was that he could not see the price of copper rising off the floor unless the whole of Chile were swallowed up by an earthquake. He had had second thoughts, though, regarding Darlington's methods. Acid processes were all very well in the British context, with skilled men and ready materials, but the colonial setting was different. He then raised a personal matter. He had been on a retainer of £300 while in Britain. Back in harness, with manifold responsibilities, he desired an increase. He retired from the room. Rymill saw that John Beck and Neville Blyth were not very happy. How much was Ayers on before? £400 came the reply, although Ayers had actually been on £40 a month. Generous Kingston thought perhaps £500 would be in order. Rymill kept quiet, lest an objection be put down to personal 'disappointment'. He confined himself to observing that responsibilities, with the new set-up, could not be as heavy as previously and any of the mine officers, or even Darlington from Britain, could manage the whole concern in addition to their present duties. In fact, because the mine was back in reasonably full production, there was quite a load for the Secretary. Correspondence was continuing to grow and meetings returned to two a month.

The contretemps between the two Henrys was developing into a limiting factor for Rymill's representation of Graham's interests. As Ayers tended to scrutinize any Rymill suggestions very closely, and as Graham's interest was significant in the concern, Rymill thought it best not to exacerbate matters by fussing over small things. Ayers had his uses, though. His being in the market for more SAMA shares would, Rymill believed, make the concern look good. However, Ayers' unwillingness to implement Darlington's acid process, at least for a while, detracted from confidence and the share price took a dive only a month later.

In mid-1871, Rymill saw another opportunity for his activism as a director. Challoner's continued presence still aggravated him. The man's squabble with Swansborough over who was 'Manager' reached the papers. Rymill would have loved to propose Challoner's dismissal but he knew he would get nowhere with it; the man had too many 'old cronies' on the board. Still, Rymill relished the idea of a dismissal, as it would force Secretary Ayers to attend to duty through monthly trips to the mine. Rymill saw his struggle on the board as a thankless task, clearly believing himself a man of ideas among all the dead wood. He was sure they were being hoodwinked – presumably by the wily Secretary – over the success, or lack of it, of Darlington's ore-dressing tower but no one would speak up. Beck and Jaffrey would praise Rymill on the quiet after a meeting but were mute while still in there. Peacock and Bullock never opened their mouths either, and Neville Blyth 'knew nothing about mining'. Kingston, as Chairman, 'pooh pooh[ed]' everything and Ayers, if he saw Rymill about to submit a written resolution, would see the meeting wound up. Once the figures were available for Darlington's ore-dressing tower, Rymill claimed that no one else could understand them until he had presented his analysis. Then, they could all see – as Ayers had already believed – that it would not do.

Rymill continued to sound very like his mentor, shoring up the relationship with Graham, now that his sister was no more. He sent family photographs, and photographs of Prospect House, as Graham had never had any of his 'Castle'. He looked forward to receiving pictures of Handschuhsheim.

Evelyn's happy delivery of Frederic Gordon was celebrated with a ball at North Terrace. Matriarch Anne invited the Rymills but they declined – 'of course'. The Ayers were well and truly back in Adelaide's social swing. Reminiscing in 1928, daughter Josey described these balls – always 'Evening Parties' – held each southern spring. The family spent a whole week preparing, washing every piece of the glass chandelier in the Large Drawing Room. The floor would be prepared for dancing by washing with milk. The two large Moreton Bay fig trees in the front garden would be festooned with coloured lamps, and the local postman, William Chapman, would play the fiddle. The family's square-piano, which normally stood in the Ballroom, doubtless gave good service, too. Poor Chapman's fingers would be taxed to the limit, with dancing continuing until all the candles had burned down, by about 4 a.m.

Governor and Lady Fergusson were doubtless invited to one of these evenings but it would be the last the couple could enjoy. This time, it was not a Governor's funeral but that of Lady Edith Fergusson which saddened the colony; she was only 32.

The Premier Question

Ayers had a few months to catch his breath before the parliamentary year commenced, in late July. Governor Fergusson was upbeat about prospects, declaring the previous year's 'depression' over. There was worldwide demand for South Australia's staple products. Whether it was public education, rail projects, free trade, or solid revenues, he was optimistic. Ayers returned to the political field rejuvenated and with reforming zeal. His first question of the Session was on the subject of wills; he saw the possibility of added revenue to be extracted from estates – perhaps a curious policy for a man of his means. He proposed William Morgan as a fitting representative at a forthcoming intercolonial conference. Morgan's delegation would have to 'appear as innocent and harmless as doves' while needing to be 'as wise as serpents' to get a good result. Was he thinking of his own nature, imagining his own heraldic birds?

The 'environment', too, was on his mind. Would the government undertake the drainage and sewerage of Adelaide, a subject of concern over the previous 20 years? The protection of native birds and animals and the preservation of oyster beds also interested him. He favoured closed seasons on these for them to recover from pollution by 'injurious matters' and did not want to see native eagles and hawks caged. He declared himself no expert on the subject, although interested in 'native' and 'noxious' species. In 1864, he had moved legislation aimed at preventing the 'Wanton Destruction of certain Wild and Acclimatized Animals'. Then, he had spoken in favour of a closed season on the killing of native and introduced species in order to prevent extinctions. Morphett, at the time, was more in favour of picking and choosing – he reckoned bald coots neither use nor ornament, as they had pecked over his hay-fields, drastically reducing the yield. Ayers' dog-control measure was not to apply to Aborigines' hunting dogs.

November brought another serious crisis in the political life of the colony over, in a sense, a trifle. The title 'Premier' was applied unofficially to the minister nominally leading government and not always, therefore, to a

Chief Secretary. Governor Fergusson tried to formalize the term, setting off a furore. He had initially induced Treasurer John Hart to take on the designation when attending an intercolonial conference. Following this precedent, he described Arthur Blyth as Premier in the November *Government Gazette*. Ayers would speak about it in the upper house, while James Penn Boucaut, one of the Members for Burra, took up the fight in the lower house. Fergusson, an ex Westminster-parliamentarian, had laid it down that the leader of a government – the Premier – should be found among the Members of the House 'nearest the people'. This was nonsense, railed Boucaut; this was not so in London, the template for South Australians' own parliamentary practice. For the moment, Gladstone or Disraeli led in the Commons, but it was quite expected that both would be succeeded by leaders in the House of Lords. Arthur Blyth defended his, and the Governor's, position. All Fergusson had wanted was that whoever led a government, from either chamber, should be regarded as Premier and, if he resigned, the whole ministry would resign with him. On this score, Ayers had a decided opinion, which he stated in the following day's debate in the upper house.

This was in the context that, to break the crisis, Blyth was advising the Governor to dissolve the House of Assembly; a Governor had no power to dissolve the upper house. Ayers rose to speak, but William Morgan jumped in first, proposing an Address to Fergusson, criticizing the choice of a Dissolution; the ministry should simply resign. In a familiar jibe, Morgan suggested that Blyth's Cabinet colleagues were not resigning as this would mean forfeiting two months' ministerial salaries. Ayers then outlined his own stance on the 'Premier' question. The term had been merely a popular title, recognizing a leader privately agreed upon between the members of the Cabinet, but certainly not bestowed by the Governor. He went beyond Morgan's position: if Governor Fergusson ignored their Address on the subject, they should take the matter directly to the Queen. They could not allow the Governor to, in effect, alter the Constitution at will. Secondly, he spoke as one who had served in eight governments and led seven of them, and he knew of no case where the leader could resign and take the whole ministry with him. He would not serve in such a government, even with the best men in South Australia. He would not place himself in that position where a 'Premier' could say: 'If you do not submit to this or that I will go to Government House and tender my resignation which is your resignation.' Citing British practice, Ayers said that though the Queen nominally chose a Prime Minister, this was obviously no longer a real power. Similarly, neither could a Governor be accorded real power to designate a 'Premier'. 'The person who formed a government should not be the be-all and end-all of it.'

Fergusson was known to have consulted Chief Justice Hanson on the question of a Dissolution and, for all the rationality of the opponents to unnecessary elections, the parliamentary stripling of an Attorney-General, Charles Mann, thought he sensed a bit of a conspiracy. He claimed that Lavington Glyde had been in touch with Ayers on the quiet, both men awaiting their chance if Blyth's government resigned. Mann reckoned this a rerun of 1868, when Hamley pulled back from a Dissolution. Fergusson, however, made of sterner stuff, stuck to his ministers' advice and dissolved the lower house. The people went to the polls only 18 months after the previous election and the Blyth team got its extra two months' salary.

A nice clipping, saved by Ayers, relates to the obvious belief of some people by the early 1860s that even MPs were getting paid. It is couched in terms of a Burra resident's sympathetic response to his MP's imagined penniless state while the parliament was not actually sitting: 'to a stump horrator of the burra minbeer of parlement, insetra insetra'. The composer, in his mock-Cornish dialect, said he was sorry that the member had no pay, with the parliament no longer sitting. He and his neighbours thought it best that his 'owl umman' [old woman] should send the MP a pasty every other day. He had got the meat for it from the 'Anti-butchers' – undoubtedly a reference to the miners' co-operative set up to supply their own needs, against the wishes of the SAMA. He would get the carrots from 'brother Robert' rather than from Tom Parks, who had taken the MP to court over a wages claim. The MP should expect the first pasty on the Wednesday, only his 'owl umman' said it would not be great as she could not get any leeks. The MP should see from all this that they had not forgotten him and the writer hoped that, when the parliament next sat, he would not forget them and raise their wages. Any ability to affect miners' wages seems to narrow the target of the spoof down to either Kingston or Ayers.

Ayers Back in Power

'Events' of one sort or another were crowding in on each other as 1871 drew towards its close. Emanuel Solomon, recently retired from the Legislative Council, was the 'hospitable and liberal' host for a dinner in the Town Hall marking the thirty-fifth anniversary of the colony. A founding member of the SAMA, Solomon was a true Old Colonist whose name was spoken in the same breath as that of Ayers and his now-knighted associates, Fisher, Morphett and Kingston.

The Governor opened the new parliament of 1872 in mid-January, speaking of much that was encouraging about the colony, though acknowledging its problems as well. With sale of government bonds well taken up in London, it appeared that the colony's credit was good once more. The people were well prepared for a greater government involvement in education – he encouraged his parliamentarians by saying that Britain had already acted in this respect, having witnessed the strides taken in Germany through Bismarck's policies.

The major problem concerned the slow progress on the telegraph line linking Adelaide with Port Darwin and, beyond, to the Mother Country itself. Legislation for the project had passed very swiftly through the parliament during Ayers' absence in Britain. It was a tricky logistical task, given the technology of the time and the difficult terrain, the wet season of the Northern Territory posing a major stumbling block. Add to these, disagreement between the leader of the project and one of Finniss' successors as Government Resident over the use of government boats, plus a heavy loss of a third of the expedition's horse power, and it is no surprise that by September 1871 only 165 miles of wire were up. The Blyth/Milne government had intervened by sending the Postmaster-General and Superintendent of Telegraphs himself, Charles Todd, to ginger up the workers, and he was given a number of additional hands for good measure. Todd could do little with the weather, though, and in March 1872 wrote to the Chief Secretary – by now a freshly reinstalled Ayers – to report the ground still too wet for his horse-teams to work.

Between the opening day of the parliament and the second sitting day, Milne had relinquished the Chief Secretaryship. The Blyth/Milne government had gone to the people, over the abstruse 'Premier' question, and the strategy had failed. It had been a costly aggravation to the colony. Rather than calling Ayers or Boucaut, Fergusson thought he had a more malleable man in H.K. Hughes in whom to entrust government. Hughes led from the position of Treasurer but took Ayers in with him as his Chief Secretary, the offending term 'Premier' passing no one's lips. The assumption generally was that Ayers was the real 'channel of communication' between Cabinet and Governor. One upper house member proposed an Address to Fergusson to clarify the position on the 'Premier' question. Ayers saw the opportunity for a little gentle chiding of the Governor by slipping in a reference at the end of the Reply. Fergusson 'noted' the comment; he had already admitted defeat and, therefore, changed his advisers. The 'Ayers-cum-Hughes' team swiftly proposed additional funds for the completion of the Adelaide–Port Darwin telegraph.

As this Session got under way, a notable voice of opposition was absent. John Baker, voted two months' absence through illness, was not to return. The Council adjourned as a mark of respect. Ayers graciously paid tribute to his 'zeal, energy, and ability'. John Barrow was frank in acknowledging that 'whatever political differences might have existed between Honorable Members and the late Mr. Baker, all . . . would be forgotten in the presence of the great leveller Death'. On that mind-concentrating note, both houses retired to see Baker laid to rest.

Rymill Emboldened

Henry Rymill was not pleased at Ayers' return to political prominence. He had had his wish in seeing Ayers' 'crony', Challoner, removed. Rymill's own man, M.H. Furniss – at Challoner's shoulder since 1852 – had been cleaning things up, as Rymill saw it. Now, just as Ayers had been settling down with the new reality, he was 'called to the ministry again'. It could only hurt the firm. He believed Ayers had already let £4,000-plus slip, through being too soft on Burra folk and their debts. Had Rymill taken on the hard-nosed mantle in the face of a softening Ayers and board, or was he just looking for ammunition?

Patience for everyone was at a premium. Fourteen days straight of 104 to 105°F in the shade meant 'young and old complain[ed] bitterly', Rymill reported. One wonders if Ayers was as happy to swelter in Adelaide's heat as he had declared himself willing to do while in a snow- and fog-bound London. Later, with a return of the very hot weather, Rymill slaved on while his family retreated to the seaside. Governor Fergusson did the same, controversially slipping away to the south-east aboard his yacht, *Lady Edith*.

Rymill took a passing shot at the outgoing Blyth/Milne government. Its members had made a mess of Northern Territory policy and now they had done the same with the overland telegraph – far from complete and costing twice the estimate. He was starting to perceive the technical innovation as nothing less than a threat to agents like himself – and therefore like Ayers, too. Graham clearly relished the development, believing the London Board of Advice could more immediately insert its oar into SAMA decision making. Rymill made it sound as though Graham would have to wait a good while for that happy day. For some, there could clearly be a downside as well as an upside to modern communications.

Ayers was soon in for more of a Rymill bashing. '[B]etween ourselves', he confided to Graham, the biggest obstacle to success was Ayers' acting as sole manager, 'carrying the directors with him on nearly everything', though shirking visits of inspection to the mine. Incomprehensible was Ayers' seeing himself as 'Secretary and Board combined . . . thoroughly doing as he likes'. Awkward as it was, Rymill claimed, he alone spoke up against Ayers.

Over the previous year, board minutes had tended to suggest a different picture – although they were compiled, of course, by Ayers himself. Coupled with Rymill's success in seeing off Challoner, it was ordered that Ayers make monthly visits. He was then 'instructed' to 'implement' recommendations from his reports. Yet, whatever the old maestro put forward as his recommendations, the younger man claimed some of *his* initiatives had produced better results just recently – if only the board were not 'lead [*sic*] by the nose as ours is'.

Ayers kept up his visitations for a couple of months and then eased off again, in the height of summer. On his own initiative, Rymill had gone up to the mine, reluctantly spending days there, considering the heat. The suggestion of a small improvement of his to Darlington's machinery, previously opposed by Ayers, was supported by the board. All Chairman Kingston could say – sounding little like the imposing 'deep bass' as heard from the Speaker's chair – was 'Then I suppose it had better be done, Ayers.' Chancing his arm further, Rymill indicated that, as there was no nominal head up at the mine, all looked to the Secretary as manager. At this, Ayers gave such a look, according to the combative young director, that no one said a word. Rymill pressed Graham to buy out the company; it could be got very reasonably. Under new management – presumably headed by one H. Rymill – the mine would pay. Broadening his attack, he took a swipe at the Ayers/Hughes ministry, before it could pass a single bill. It would be better for the colony if the democratic experiment were ended and a return made to the nominee-legislature of old.

Whatever Rymill was, or believed he was, accomplishing for the benefit of the mine, his efforts were going unrewarded by the international copper market, which discouraged him by its fall the following month. He groped for further accusations against Ayers, depicting him as captive of a regime of primitive industrial methods that had advantaged the E&ACC all these years. The board was astonished at this 'revelation' but Ayers dismissed the criticism. Well he might have. Had he not fought the good fight against Burr's assaying failures? These were followed by the long-running tussle over agreeing ore qualities between the SAMA and the E&ACC. Through James Coke and his Swansea principals, Ayers had all the practical explanations he needed – diagrams are still extant. Ayers had once frankly admitted to the E&ACC's James Hamilton that disagreements over assays were a standard game they played, with compromises always happily agreed in the end. Whatever Rymill believed, Ayers could not exactly be considered a novice in the field. Rymill's contention to Graham that the Burra had really worked itself, Ayers merely looking after the money, was equally unsupportable. Rymill won a small victory regarding his criticism about the lack of a paved ore floor, for clean assays, as the board told Ayers to change the practice. Rymill believed a new manager, being sent over by Graham, was more than ever needed as Ayers had still not been to the mine, nor looked like going. How could the concern 'thrive without a head?' The tone of the forthcoming Annual Report was not too bad, Rymill thought, and was largely the work of Mr Furniss, though his name would not appear anywhere near it. That was 'frequently the way a discerning officer gets treated [by Ayers] whereas a thoroughly useless one like Challoner was could not be made too much of'.

With the AGM coming up, Rymill sought new board blood that would give him support, and saw some potential in Lavington Glyde. The election of directors, however, went against Rymill. Ayers' friend, Samuel Tomkinson, instead joined the board. Tomkinson, like Furniss, had been in his job – manager of the Bank of Australasia – since 1852 and, as the SAMA's banker, he naturally had a long history with Ayers. Glyde had refused to stand, saying that he would 'not oppose an old friend' – whether he meant Ayers or Tomkinson is not clear. So, the board was 'still the same "Happy family"'. One more line of attack remained, with Rymill asking Neville Blyth and John Beck back to his room in Imperial Chambers; could Ayers be induced to take a simple directorship and allow a manager to go and run things at the mine? Beck was for but Blyth against, so that would not happen. He was out of ideas.

Graham's impatience for the telegraph grew, but 'remote control' would not improve matters, Rymill argued. The current board's ignorance, coupled with Ayers' unwillingness to 'spend a shilling unless he [could] see a certain return of 2' compounded Ayers' lack of knowledge of the industry. With Tomkinson at the board, Rymill had 'pretty daringly (perhaps I should say very)' brought up the way matters were being allowed to drift. Ayers must be made to visit monthly again. In view of the many suggestions coming from Mr Rymill, Ayers thought perhaps a deputation of the directors should go up with him. For economy's sake, the directors decided that Ayers and Rymill should go, which made Rymill very uncomfortable. He had no alternative but to agree, he told Graham, because he represented the latter's large interest. John Beck decided to go with them and Rymill expected that more good would come out of this particular visit than any previous director-visit in the mine's history – a big claim.

Uncomfortable about the journey as the younger Henry was, the older was clearly not put out in any way, being 'extremely chatty and agreeable and quite willing to renew our former intimate acquaintance – but this can never be as far as I am concerned'. A concession emerged following the visit: Ayers suggested a meeting to formulate a joint report prior to the next board meeting. Although Rymill considered this an attempt to avoid raising conflicting ideas before the full board, he derived some satisfaction from the novelty. He must have felt he was wearing Ayers down.

16

TELEGRAPHS, HONOURS AND THE HARD SLOG OF GOVERNMENT ONCE MORE

Telegraph greetings from London finally reached Adelaide, in November 1872, thanks to Charles Todd's energy and 15-year obsession with the project. He had worked away through the Northern Territory's dry season, driving ever southward and completing the line – known to the Marryat Creek Aborigines as 'whitefellow wheelbarrow curtejabba' – by August. 15 November saw great celebrations, Todd and his men processing behind a brass band. Celebratory games included enthusiastic 'kiss-in-the-ring' sport for some young people. At an ox roast, with numbers taken to the hospital with sunstroke, Ayers praised the men's achievement. They did not know him, but he knew their task intimately, and their hardships, perils and zeal. In a barely cooler Town Hall, Todd had his apparatus set up as a backdrop to the obligatory banquet. The dreadful acoustic drowned the speech-making but tap-tapping down the wire came news. Todd received a CMG. Governor Fergusson would be posted to New Zealand. Ayers, at the head of government, received a Knighthood in the Order of St Michael and St George. The diners rose and gave long applause. Ayers was the first member of the legislature with a knighthood in one of the Orders of Chivalry. As with Oscar winners, he accepted, certain that it was not just meant for him personally but as a mark of honour for the colony. Sir Henry kept a poetic outburst from some wag, in a scrapbook, one verse running:

> Oh! if I were a K.C.M.G.,
> Like good Sir Henry Ayers,
> My life would one long summer be,
> With no desires or prayers.
> I'd scorn contentment's lonely dells,
> I'd scale ambition's heights,
> I'd only mingle with the swells
> And laugh at paltry knights.

Months before the honour was granted, a sour letter to the *Register* saw it coming:

> HOW TO GET KNIGHTED
> Sir – The process is quite simple. First, you get into one Ministry by abandoning your principles. Secondly, you get into another by abandoning all your colleagues. Thirdly, you take credit for work performed by other men for opening a line of telegraph, with the construction of which you have not had the slightest connection. Having thus proved yourself one of Nature's noblemen, is it to be wondered at that her Gracious Majesty recognises the fact? The above is respectfully submitted to members of Parliament, for their guidance and imitation.

The author, 'B', could be Boucaut, dropped from government a day or two before. We might have expected Henry Rymill to say something begrudging about this Honour, too, but he was surprisingly accepting. He hoped, in vain, that it would be 'in fra – dig' for a knight of the realm to continue in the position of a company Secretary.

Some optimists floated a company to build a transcontinental railway along the telegraph line. Numbers of Sir Henry's associates bought in but, significantly, the experienced financier held aloof. Rymill, having learned good lessons from Ayers, warned Graham that buying in would be madness. Some still thought so in the twenty-first century, when a northward rail extension from Alice Springs finally linked Adelaide with Darwin.

Pushing for Reform

Before turning enthusiastically to reform, Ayers needed to stabilize his ministry. Attorney-General Boucaut had to step down while fighting allegations of corruption over the long-running Moonta Mines case, forcing a reshuffle. Boucaut begged and got Ayers' backing and was indeed cleared. Hughes'

Treasury position was taken over by Barrow, owner of the *Advertiser*, founded as an 'anti-squatter' newspaper in 1858, with Ayers chairing its board. There were probably enough independently minded MPs to support the new Ayers-cum-Barrow ministry, thought the *Register*, continuing:

> His [Ayers'] business ability, his ready tact, unimpeachable integrity, and his political experience, were all brought forward, if not as reasons for his continuance in office, at all events as proofs that he had not forfeited the confidence of the country by being associated with the men who were then his colleagues.

While the editor thought the four dropped ministers had perhaps 'not been handsomely treated', the Governor had done what he thought best and Ayers had constructed what *he* felt was a strong ministry. A letter to the editor, from 'Phoenix', gave the new team its nickname, the 'C.R.A.B.S.' ministry, from the initials of its members' names. One or two of its members 'inherit[ed] a large portion of the peculiar nature of their crustaceous namesakes'.

A small but significant reform satisfied Barrow's ambition in simplifying the dense legalese of legislation. In Britain, it was to the upper house that difficult legal matters were sent, as the Law Lords sat there. In the South Australian parliament, legal minds tended to be found in the Lower. However, as William Milne observed with a twinkle, their own Mr Ayers had an opinion on every legal matter that came before the Council.

Two major reforms interested Ayers. He had talked of state education in his 1857 election platform. Now he brought it before the parliament, drawing upon work in Canada and the United States, having also witnessed moves in that direction while in Britain. The main hurdle for such a programme was religion; no denomination wanted the doctrines of another pushed down their children's throats. Hence the attraction of a secular system to Ayers and people like him. Others feared the outcome would be an educated but godless society, storing up problems for the country's future. Still others harboured outright opposition to the notion of public education, although this was an untenable position; the colony was becoming concerned at the strides being made by European countries embracing education for their populations. The Ayers proposals were not to come free. Parents would have to contribute a small amount, but the government would fund a school-building programme and provide the bulk of the support. Given the division of opinions, the measure failed to get a majority.

More immediately successful was Ayers' embrace of policies bringing fundamental change to the treatment of children of the 'destitute poor'. With

Adelaide's two institutions full to overflowing, children were already being fostered out but with no supervision of conditions. Ayers was adamant that Britain's way of dealing with its destitute – admittedly on a far greater scale – was utterly unsuited to the South Australian context. Yes, he wanted to avoid anything that might encourage pauperism (long-term welfare dependence), especially perpetuating it down the generations. What he did not want to do, though, was to stigmatize the children. Many women and children were unable to fend for themselves through accident. Their main support might have deserted them, be ill, dead or a drunkard. While for children convicted of a crime there could be a Reform School, for the simply destitute, he advocated an Industrial School teaching useful skills. At the same time, they could be boarded out, living with willing families who could nurture them in an encouraging environment.

In this, he was fully in tune with the principles of the Boarding-Out Society, formed that year by members of the Unitarian chapel in Wakefield Street. Ayers had shifted his allegiance to this chapel by 1865, at the latest. It was home to a small intellectual and cultural elite, notably including Catherine Helen Spence – thinker, writer and lecturer on many reform topics, her name later linked to women's right to vote. Perhaps Ayers had sought something more meaningful than his nominal Anglicanism following the profound experience of losing baby Sydney. The progressive tenets of the congregation – discounting any evidence for the belief in the Trinity, seeing Jesus more as a son of God than God's Son – clearly appealed to Ayers, as witnessed by his wry view in Britain of escaping a sermon on the Trinity and his praise for the Jewish faith as bringing 'intelligence' into the worship of a deity. Given Ayers' stated faith in science, exemplified by his desire to meet T.H. Huxley while in London, his views would also chime well with the explicit statement of an exchange preacher in the Wakefield Street chapel shortly after that: 'God speaks through the voice of science today as he spoke by the prophets of old'; it was God who had instigated evolution. Not the least noteworthy point was the fact that these assertions were made by a female visiting preacher. The lighter side of this progressive attitude is captured in an extract from the *New Science Bible* which Ayers kept in a scrapbook. It parodies the words of Genesis, seen through a science prism:

Contents – Atoms, Apes and Men
1. Primarily, the Unknowable moved upon the cosmos and evolved protoplasm.
2. And protoplasm was inorganic and undifferentiated, and a spirit of evolution moved upon the fluid mass.

3. And the Unknowable said, Let atoms attract; and their contact begat light, heat and electricity.
   Etc.

Allowing for John Baker, Ayers was very much among friends in this small Unitarian congregation. His brother-in-law, A.J. Baker, Williams Sanders and Morgan, and numbers of his business and political associates, held seats there. The intellectual culture, not to mention the enthusiasm of the chapel founders, the Clarks of Hazelwood Park, for amateur dramatics, would have held great appeal for Ayers. John Howard Clark was both a SAMA shareholder and an auditor of the Savings Bank, with which Ayers was so long connected. The arts were represented by the photographer, Duryea, and Ayers' postman/fiddle-player, William Chapman.

The Unitarian representation among the initial supporters of the Boarding-Out Society was strong: Anne Ayers was on the committee with Misses Clark and Spence, and Maggie Ayers was one of the 18 early subscribers, together with her father and mother. The ladies gave of their time to inspect the homes of the prosperous working class into which the destitute children were placed. It was intended that the visits should be irregular but reasonably frequent.

Ayers was insistent that under no circumstances should it ever be thrown up to the children that they were being rescued from pauperism. He was glad, he said, that there were no large charitable institutions in the colony doing this sort of work, as in Britain. It was better done by government, he believed. Doubtless remembering conversations on charity at the Leightons' country-house in Shropshire, he said he did not want 'Bread Sundays' and the like. His further appreciation of the way London's Jewish community handled the problem may well have played its part in establishing his own position.

Someone at the Port was offering to take charge of 40-odd boys, and turn them into seamen, which was commendable, but Ayers did not want to see this before age 12. Maybe his own experience of entering work at 11, or Frank Potts' tales of the Navy at nine were salutary, or perhaps times had simply changed. If there was any fear of mistreatment, Ayers insisted on very close scrutiny. The poor 'would always be with them' so it was the community's duty – as expressed through the parliament – to support them.

William Morgan was happy to support Ayers in the Boarding Out measure – he was another Unitarian adherent after all – and regular Ayers supporter William Parkin spoke of the obvious breadth of reading and research which had underpinned Ayers' moving of the Bill.

Another issue Ayers put through the Legislative Council was a legal one, to be sure, but squarely in the arena of women's rights: a Married Women's

Property Act. This was designed to change the situation whereby, on marriage, a woman's property became that of her husband. Ayers described the reform as a 'simple act of justice to women'. He could not understand why 'in this day and age' it had not already been enacted. These were rights, he believed, that should always have been there. The only possible danger he could see, which would have to be carefully watched, was that some tradesmen might transfer their assets to their wives to avoid their creditors. The bill did not go as far as John Stuart Mill advocated in giving the right to vote to women, but he felt it would add to the 'future political and social importance of married women' who had been 'in a lower position than was right for centuries'. Ten years before, a Melbourne paper had ridiculed Port Adelaide for allowing women to vote for its Council. The Property Bill went to the lower house but failed to pass before the end of the Session, and so was lost. Nevertheless, Ayers had placed women's rights on the South Australian agenda.

Pursuing his concern of the previous year, Ayers' government introduced environmental reforms. A 'Protection of Game Act' was put through to prevent the 'wanton destruction' of oyster beds, and for the protection of some birds, and other animals. Native animals were all included, and the 'game' mentioned included imports like partridge and grouse, which were to be spared the guns for a two-year establishment period. Having already had a bad experience with rabbits, it was felt that importing hares would be irresponsible. The protection Act had its critics, oyster beds representing a role-of-government issue for some. A.B. Murray, for instance, believed bed-owners should regulate themselves. In which case, Ayers riposted, do away with government, and the police force as well. Murray, unconvinced, said it was like telling pastoralists how and when they should breed their stock. A poorly chosen argument, thought the Chief Secretary, as governments did that already, for example, to prevent scabby sheep. Murray was out of arguments.

Numbers of other issues were addressed by the Ayers government. The telegraph had brought the issue of intellectual property to prominence, requiring legislation concerning copyright of telegrams. Primogeniture was abandoned as inappropriate in a growing colony, and imprisonment for debt was ended. Licensing laws were tightened up, with pub landlords made responsible for drunk and disorderly behaviour, gaming or prostitution on their premises. Ayers would publicly declare himself in favour of Temperance; he could hardly have been a teetotaller, considering what he put away, both personally and in his cellars, during the London trip. Corporal punishment for wayward youth was restricted to misdemeanours far more serious than breaking a window – or possibly even more so than the activities of 'unruly half-grown boys' noted in the Botanic Gardens, in drawing moustaches on the statuary and scrawling 'abominable and indecent sentences on the pedestals'.

Despite the failure of his public schooling legislation, education was on Ayers' mind in another respect through the latter part of 1872. A strong movement had arisen in favour of converting the Union theological college into something grander: a university. Ayers' name went forward as a member of the University Association. As Chief Secretary, it was to him that its committee came when it had concrete proposals necessitating legislation. Ayers' government wanted to see evidence of good financial backing from the public, in addition to the very large promised donation of £20,000 from Walter Watson Hughes. There had also to be an unequivocal statement that the institution would have no truck with sectarianism or denominationalism. Then, a university act could be put through the parliament and a grant of at least four acres, on North Terrace, would be forthcoming. The Chief Secretary was approached for a personal donation, and did not disappoint. Along with that of Peter Prankerd, Ayers' contribution of £500 was the next largest after W.W. Hughes – at least until Thomas Elder came forward and matched Hughes' a couple of years later. Even Governor Fergusson's gift was only a fifth of Ayers', and R.I. Stow's and others' only equalled a tenth, while most gifts were the one- and two-guinea varieties.

The time came for activist Governor Fergusson to close the parliamentary Session, before leaving for New Zealand. Sir Henry's chum, gruff Speaker Kingston, took the opportunity for exacting a little revenge. Fergusson and Kingston had clashed over one of those abstruse Parliamentary Privilege questions, and Ayers had put a placatory blanket over it with one of his classic reasoned parliamentary speeches. On the last day of parliament, Speaker Kingston went up to Government House to obtain Fergusson's assent to the most recent Act. Unfortunately, he had the paperwork of another act caught up in it and the Governor overreacted. This rekindled the heat of the previous Privilege controversy. When the lower-house members received the traditional invitation to appear at the bar of the Legislative Council to witness the closing of the Session, Kingston had the Assembly doors not just shut, but locked in the face of the Sergeant-at-Arms, the lower house continuing its debate. In the Council, Sir Henry and his colleagues were nonplussed while a fuming Fergusson had a steamer to catch. Barrow tried to calm the lower house. Messrs Mortlock and Riddoch whipped round the outside and broke glass in reopening the doors – dark memories of the Commons in 1629 came to some minds! With most of his parliamentarians assembled, Fergusson resumed his position, closing the Session as though nothing had happened.

Harry Ayers again deputized at the SAMA in late 1872 when Sir Henry sailed for an intercolonial conference in Sydney. Treasurer Barrow was his colleague and Ernest Ayers went along for the ride. The visitors spent three weeks in the New South Wales capital, the conference delegates receiving a warm welcome and generous treatment. The perennial problems with the sea Mails were of main interest for South Australia, even though the P&O steamers were at last calling at Glenelg. Sir Henry, in a speech to the delegates, was full of praise for the host colony. If he were not a proud South Australian, he should like to have been a New South Welshman. He went on to look forward to a union of the colonies. If he could see a 'one and undivided dominion' of Australia, he could die happy. Many questioned the value of these conferences, but he felt that the opportunity they afforded for ministers to talk with people they previously only knew through correspondence was more valuable than could ever be represented on the printed page. He sounded a convinced federalist.

Mining Momentum

Back in Adelaide, Ayers immediately became a grandfather again, to Harry and Ada's Amy Josephine. His return also marked the end of Swansborough's oversight of the Burra. According to Henry Rymill, Swansborough had never got the Burra properly 'under his grip', unlike the renowned Captain Hancock of the Moonta mine. Where Darlington's crushing machine was inefficient and rarely working, Hancock's 'jigger', recently installed at Burra, was making light work of the old mine's produce. Rymill played down Ayers' role in any innovation to Graham, reasserting that numbers of the directors were grateful to him (Rymill), on the quiet, for his energetic pursuit of efficiencies and profit for the concern.

As luck would have it, a rise in the world copper price coincided with the cost savings at the old mine. Even orthodox underground work was producing profitably and a dividend of 50 per cent was managed, the first since 1864. With shares rising somewhat, some old hands were seeking to buy, and Rymill was pretty sure that some bids came from the Secretary himself. He was surprised at Ayers, whom he believed had never

> realised the true position of this splendid property in its' [sic] altered condition until I pointed out to him, as fully as I could, that it merely needed ordinary attention and proper working to make it a good investment and how much I should like ten or twenty shares in it.

Rymill congratulated himself that he had Graham, in Europe, better informed as to the state of things than some, he said, who ought to know better. Rymill induced other local investors to buy, hoping it might 'lend a beneficial effect upon "him" [Ayers] to keep him up to the mark in working the property in a manner it should be and which its importance deserves'. Increased momentum warranted weekly board meetings, agreed to 'after some little hesitation on the part of Ayers'. Some thought the shares would go to £100 again, but Sir Henry had already bailed out at £65. While bringing good tidings to Graham, Rymill thought he might broach the subject of a bonus for the good work he was also doing in the expansion of the Canowie sheep station, which Graham quarter owned. Graham had agreed to the expansion by letter, but then countermanded it by telegraph. Rymill had feared the telegraph, and his bonus mirage vanished down the wires.

A Robert Sanders – no relation to Frank Rymill's father-in-law, Billy – took over from Swansborough. Rymill, suddenly less decisive, got the jitters about the changeover, although he seemed a 'likely man . . . a pushing man' who considered the '£.S.D. question'. The enthusiastic director soon came round. Doing away with the skills of the deep-mining Cornish elite, Sanders instituted an expansion of the open-cut technique. Before too long, in the hope of further reducing manpower, Sir Henry canvassed the introduction of new-fangled machinery, such as pneumatic drills.

Given problems over the will of 'the Mother', recently deceased, it suddenly struck Rymill to ask Graham who were *his* executors – he believed Ayers had once been named, though he could not imagine that had continued 'after what [had] come to your knowledge'. Rymill could only offer to work with whomever Graham had named 'without any unpleasantness' – though he hoped it would be 'years yet'.

Facing the Voters

In office, and approaching 52, Sir Henry considered his political future. Far from any suggestions about giving up public life, he faced the voters again in 1873. As before, he contented himself with a brief 'pitch', little more than a statement of his candidacy and a claim on the electors' support if they valued the fact that he had 'never hesitated to give utterance to [his] opinions on the various subjects' raised in the Legislative Council during the previous 16 years. There was a spirited fight over the one lower-house seat being contested – that of Burra – but, over the seven seats in the Legislative Council, there was clearly voter apathy, with a low turnout. Little work, then, for Harry in scrutinizing the vote-count on his father's behalf. Sir Henry, at any rate, outshone all others to head the poll a second time.

Facing SAMA voters was a hit-or-miss affair for Henry Rymill at the same period. Either he or John Bullock would have to step down through the drawing of lots. In the event, Rymill managed to keep this tilt-yard open for his jousts with Sir Henry by persuading Bullock to retire. In Rymill's opinion the man only came 'when there [was] something to be got out of it'. Rymill topped his own poll with 170 votes.

Ayers' personal happiness was also increased, between the election and the first Sitting of parliament, by another addition to the family. In early June, Fred and Evelyn produced another little boy, Julian.

Following Ayers' re-election, the *Register* chided him for slowness in getting parliament sitting. His public education measure was on hold, pending observation of Victoria's experiment in that sphere. The colony had turned the corner and revenues were rolling in to the Treasury; this was the time for good social reforms, urged the newspaperman. Ayers' government was also indifferent over Civil Service reform, the criticism continued. All Sir Henry's government did was quietly and efficiently administrate; there were no bold reforming measures in the offing.

In the meantime, those who could remember the Adelaide of 1851 had some sense of déjà vu. Shop assistants and others took ship for new gold-fields, in South Australia's own colony, the Northern Territory, from where the siren call of the Yam Creek gold reef could be heard. At least, this gold 'rush' was nothing like as frenetic as the previous one.

With the new Session of parliament imminent, Ayers received a deputation of farmers, desperate through a labour shortage. Sir George Kingston ushered them in to the presence. Sir Henry welcomed them warmly, telling them that the government was similarly short of workers for its projects. Their problems were exacerbated by a 'bounty of Providence', and the colony's thriving state. He had been asked, what was Responsible Government? Government in South Australia had wide responsibilities, he continued; it provided railways to carry their grain products from a very extensive area and, even in the recess, the government had gone as far as it believed permissible in providing more rolling stock. He reminded them that he had previously met them, in 1868, when there was a shortage of seed wheat, and that there was no greater supporter of the wheat farmer, as exporter, than himself. Everything that now could be done would be done, but they could not expect five men (the ministry) to produce labour just when they wanted it, in the same way that seed wheat could not be magic-ed at the time the Almighty had deprived them of it. There were always other interests to balance. With still a week remaining before parliament sat, he could not expend £100,000 without its say-so. If parliament gave the word, he would telegraph for an influx of immigrants immediately. What if he had done so earlier and a dry summer had ensued? Yet, he had always supported immigration, he said – it would have been hypocritical for him to have been against it! It sent a good message Home, that people could come, work and hope to afford a property of their own. Major-domo Kingston ushered out the farmers, thanking Sir Henry for kindly receiving them.

Immigration proved Sir Henry's downfall. 'Tea-pot' Tommy Reynolds, one of the government's strongest members, walked out once again. For 20 years he had opposed expenditure on immigration. Rumour had it that this was not the whole story, and that Reynolds was eyeing the post of Resident Administrator in Palmerston (Port Darwin). The *Register* editor could not credit Sir Henry would sanction such a move, which would sink the ministry and mean political suicide – not one of the Chief Secretary's failings. Sir Henry was believed to have spent an entire Saturday trying to persuade Arthur Blyth to join his ministry, to no avail. Unwilling to continue with a makeshift Cabinet, Sir Henry had no alternative but to resign, a couple of days before parliament sat. The *Register* praised Sir Henry's wisdom and self-denial in not clinging on to power. It praised the achievements his government could claim and, even if it had not met expectations, those had been set ridiculously high. It had passed useful if flawed legislation, and Chief Secretary and Treasurer had had success at the Sydney conference, in agreeing border duties with New South Wales.

One of Sir Henry's last acts had been to welcome the new Governor, Musgrave, arriving in a colony given a poor review by a visiting Anthony Trollope. At least his mention of Sir Henry in the write-up was not adverse. The press retaliated with the criticism that Trollope's work was not up to the standard of either Thackeray on Ireland, or Dickens on America.

Into the leadership void stepped Arthur Blyth, bringing Lavington Glyde with him into the ministry. Henry Rymill noted Ayers' downfall without crowing too much. The parliamentary scene was changing. Sir John Morphett retired and William Milne accepted the Presidency (Speakership) of the Legislative Council. It was acknowledged that Sir Henry would have been first choice but, still vigorous, he relished the cut and thrust of the chamber floor. There was still much to do.

17

THE ROCK

Sat on a camel for the first time in his life, William Christie Gosse set out from 'The Alice Springs' late in April. He led the 'Central and Western Exploration Expedition, 1873', with his party of four whites, including his brother, a native boy, Moses, and three 'Afghan' cameleers, most prominent of whom was Kamran. Instructions from Ayers' government were to make for Perth in as straight a line as possible from the transcontinental telegraph line, avoiding crossing paths with a privately funded expedition heading much the same way. Gosse's party crossed and recrossed the line of Ernest Giles' earlier expedition and then made its way across the eastern arm of Lake Amadeus. From there, Gosse says, they went on to a 'high hill, east of Mount Olga [Kata Tjuta], which I named Ayers Rock'. He gave this accolade from the centre of the massive island-continent, just days before Sir Henry lost office.

The land crossed was unpromising, consisting of a few small mulga (acacia) patches, but 'nothing fit for occupation'. '[G]ood country' – of pastoral interest – was rendered valueless because access was impracticable. Waters previously thought permanent had clearly been dry for months. The spring at Uluru was the first water source he had found in 80 miles.

His diary reveals how profoundly the extraordinary feature affected him. With images of the monolith now so familiar worldwide, it is difficult to comprehend the impression conjured up almost exclusively by Gosse's word-painting as only one drawing appears to have come out of the expedition – done by second-in-command, Edwin Berry. Reading Gosse's diary, we may capture something of the picture Sir Henry would have formed when he pored over the official report months after the discovery.

> Sunday July 20, – Ayers Rock. Barometer 28.07 in., wind east. I rode round the foot of the rock in search of a place to ascend; found a waterhole on south side, near which I made an attempt to reach the top, but found it hopeless. Continued along to the west, and discovered a strong spring coming from the centre of the rock, and pouring down some very deep gullies into a large deep hole at the foot of rock. This I have named Maggie's Spring [Muṯitjulu Waterhole]. Seeing a spur less abrupt than the rest of the rock, I left the camels here, and after walking and scrambling two miles barefooted, over sharp rocks, succeeded in reaching the summit, and had a view that repaid me for my trouble – Kamran accompanied me. The top is covered with small holes in the rock, varying in size from two to twelve feet diameter, all partly filled with water.

Gosse, the first white man to reach Uluṟu, was the first of many to climb the Rock – a controversial act given its spiritual significance for the local Aboriginal people. From the top he describes surrounding landmarks:

> How I envy Kamran his hard feet; he seemed to enjoy the walking about with bare feet, while mine were all in blisters, and it was as much as I could do to stand; the soil around the rock is rich and black. This seems to be a favorite resort for the natives in the wet season, judging from the numerous camps in every cave. These caves are formed by large pieces breaking off the main rock and falling to the foot. The blacks make holes under them, and the heat of their fires causes the rock to shell off, forming large arches. They amuse themselves covering these with all sorts of devices – some of snakes, very cleverly done, others of two hearts joined together; and in one I noticed a drawing of a creek, with an emu track going along the centre. I shall have more time to examine these when the main camp is here. This rock is certainly the most wonderful natural feature I have ever seen. What a grand sight this must present in the wet season; waterfalls in every direction. I shall start back, to morrow, and trust to finding some water between here and King's Creek, which is now eighty-four miles distant.

These were Gosse's immediate reactions to the monolith which gave the experienced explorer such a unique experience. He could fittingly have named it to honour a member of the royal family or the newly arrived Governor, Anthony Musgrave. Gosse gave a nearby range of hills Musgrave's name.

Was it just Ayers' good luck to get his name made famous, worldwide, by happening to be 'Premier' when Gosse's expedition was despatched? In that case, we might easily know instead of Bagot Rock, Strangways, Hart, Blyth or Milne Rock. Rather, there appears to be a little more to this naming.

True, the name honoured a respected South Australian personality but, beyond that, it was a tribute to a family friend. The younger William Gosse's father, a Hampshire man like Ayers, was a notable Adelaide medical man. Living, like Ayers, on North Terrace, he did respectable things, such as sitting upon the board of the Institute, furthering the course of knowledge and personal improvement in the population. He may well have been the Ayers' family doctor. Ayers and Dr Gosse had shared a Sunday-morning gossip in London, but this was no more than their regular routine at home, over a glass of sherry. It is no coincidence that the first fresh spring of water that Gosse came across he named Maggie's. The explorer carried a candle for Maggie Ayers, six years his junior and unmarried. Gosse had known personal tragedy: his young Melbourne-born wife of only five months had died of typhoid, in 1868. He had ridden 70 miles from his survey camp to be at her death-bed, his father unable to save her. Five years on, he clearly felt ready to embark on a second marriage; he had many hours' opportunity to contemplate his future, out in the wilds. The monolith-naming was undoubtedly intended to stand the young explorer high enough in Sir Henry's good books for the knight to look favourably upon a proposal. It was a down-payment of sorts for Maggie's hand, backed up shortly with a 'Harry's Reservoir'.

Gosse's fascination with the Rock continued, once he had set up camp there:

Monday, July 28, – Depôt No. 8, Ayers Rock. Barometer 28 in.: latitude 25° 21' 28" south. This Rock appears more wonderful every time I look at it, and I may say it is a sight worth riding over eighty-four miles of spinifex sandhills to see.

He had speculated about the stunning appearance the Rock would present in the wet season and though he was there in the drier season, he was lucky enough to witness what he had imagined:

Friday, August 1, – Depôt No. 8, Ayers Rock. Barometer 27.87 in.; wind N.W. The Rock presented a grand appearance, this morning; close to our camp [on the south side, about half a mile from Maggie's Spring] was a waterfall about 200 feet high, the water coming down in one sheet of foam.

The rainfall may have been the accidental cause of an interaction of Gosse's party with the Aboriginal people of the area. On the other hand, it may just have been that the locals had had sufficient time to observe the intruders to venture a cautious meeting:

> Sunday, August 3, – Depôt 8, Ayers Rock. Barometer 28.03 in.; wind S.W. A beautiful day after the rain. Walking about the rock on the west side I observed native fires quite close to us, and soon after two natives came for water, and after our making signs, they came up to us, but seemed terribly frightened. I fancy they must have heard of whites before. They were all fine looking young men, about 5'8" high, wearing their hair in the shape of a chignon, a string being tied tight, close to the head, the same as those on the Finke [River] and on the telegraph line. I took advantage of this fine day to kill a bullock.

Gosse had picked the wrong time for drying, or 'jerking', the meat as the weather turned wet again. While he struggled with the conditions, three more Aborigines arrived. They were 'very peaceable' and he gave them a fire stick. The only word of their language he managed to gather was 'carpee', their word for water. Perhaps extraordinarily, he makes no mention of whether he tried to find out their name for the Rock.

On 8 August, Gosse headed off west but returned having found very little promising pastoral land. On reaching Charlotte Springs, he sent a preliminary notice of what had been accomplished, which appeared in the *Register* on 22 December. The newspaper advocated sympathy for Gosse's inability to complete his full mission into Western Australia, for lack of water. Shortly, a slightly fuller report spoke of the 'romantic discovery' of the Rock. It was reckoned a most notable achievement, finding this 'huge monolith', 1,160 feet high and six or seven miles around the base, in this 'out of the way place'. The romantic aspect was believed increased by the fact that 'a stream of water fed by a spring in the centre of the conglomerate flows from the rock'. Those interested in geology would impatiently await further news of what was 'destined to become famous as one of the most singular and most prominent of Australian landmarks'.

Ernest Giles, leading another private expedition, might have wished that he had paid more attention to the monolith when he had spotted it in the distance the year before. He had already honoured Ayers, independently, naming a peak in his 'Ayers Range' Mount Sir Henry. Giles appears less adept at dealing with encounters with Aborigines, resorting too readily to warning shots and, unlike Gosse, appears less aware of their general presence. Western Australia took its toll on his party and he returned via 'Ayers Rock'.

While Giles was clearly put out by the number of times he came across Gosse's tracks or camp-sites – apologizing in his journal if his comments sounded sometimes overly 'warm' – he too was entranced, and seems to half offer an excuse for more or less discounting the Rock previously. He had failed to comprehend its size and thought it 'too small to engage much attention'. 'Mount Olga', he went on, 'is the more wonderful and grotesque, Mount Ayers the more ancient and sublime'. It was 'the most singular looking mount'.

Nothing was said regarding the Rock's naming in the *Register*'s initial reports, though perhaps the excited explorer beetled along North Terrace, at the earliest opportunity, to give a personal report to the dedicatee. Otherwise, it might not have been until Gosse's official report that Sir Henry got some feel for the extraordinary landmark with which his name had become linked, and which, indeed, would come to overshadow memories of the man himself.

Gosse had clearly already known of a rival for Maggie's heart, applying the name of the young man to another feature in the vicinity, Lungley's Gully. By the time he returned from his expedition, it was already clear, according to Rymill, that Maggie's heart was inclining in the Lungley direction. Gosse had to look elsewhere and married Agnes, the daughter of one of Sir Henry's Legislative Council colleagues, Alexander Hay. The widower, Hay, like Sir Henry a one-time supported-emigrant-made-good, had married Gosse's older sister – also an Agnes – two years before, so the relationships between the Gosses and Hays became highly convoluted. Highlighting the 20-year age difference between their sister and bridegroom Alexander, the Gosse brothers had put up the not too kindly farm sign: 'OLD HAY FOR SALE'.

Industrial Harmony?

As Gosse was returning from his explorations, Sir Henry demonstrated what strides he had made over the years in transforming himself from the hard-driving young manager who had outfaced the striking Burra workforce back in 1848. He was guest speaker at an Adelaide dinner given by employees for their employers, celebrating an anniversary of Victoria's introduction of the Eight-Hour (working) Day. The principle of eight hours' work, eight hours' recreation and eight of rest, was one Sir Henry very much favoured. News of his speech as reported in Britain, in an article headed 'From the Colonies', came bouncing back to South Australia. The British reporter gave a glowing testimonial to the harmony and good feeling exhibited between employers and employed in the colony, and to the remarkable quality of Sir Henry's speech. Though brief, it showed 'such a marked ability of thinking, so wide and statesmanlike a range . . . sound and scientific accuracy in a vexed question'.

The reporter reckoned such a speech rare in England when 'politicians and economists express their views'. It was good to see employers and employees recognizing that they were partners in a common enterprise, the writer noted, not then prevalent at Home, and both accepting the benefits of the Eight-Hour Day. He feared such would be a long way off for British workers. Sir Henry had spoken highly of the benefits for workers that had come from steam and machinery. Recent developments had made it possible to provide more liberally for their families, both in necessities and luxuries. 'Amusement', said the reporter, 'was preached as beautiful', and Sir Henry advised workers to relax in ways that differed from their line of work. He had admonished Frank and Harry, from London, against 'All work and no play'. Sir Henry had clearly become an advocate for work–life balance, perhaps reflecting that he himself had this wrong in the past. Workers should in their leisure time, however, educate themselves, he said, and keep up with public developments so that they could 'attend to the quality of the men they elect[ed]'. They had to think for the 'whole society, not just a class interest'. The reporter reckoned all this a tribute to Sir Henry and the people of South Australia that such things could be said so frankly. The report is so effusive that one even wonders if Sir Henry had some hand in drafting it!

Industrial harmony was conspicuously absent from one of the last SAMA Board meetings that 83-year-old William Peacock was able to attend. If Challoner had been Sir Henry's creature in the mine office, Furniss was Rymill's. Beck put up a resolution to prevent these two communicating privately. Rymill, seeing the resolution was in Kingston's handwriting, was sure Ayers lay behind it. Fearing the 'Trio' ganging up against him, Rymill was 'rattled', he admitted to Graham. With Sanders proving his worth as mine Captain, Rymill felt his own services no longer valued. He made a stand against Beck's resolution and launched into a thorough recital of past mismanagement and

> want of interest by every officer as well & Directors connected with the Company in its affairs & instead of its being made one of the largest undertakings here it was treated as nothing else than a chandlers shop & the very thing they now objected to had in consequence been forced upon me from the absence of any control being brought to bear on our doings at the mine & I pointed out that as far back as when I moved the resolution for the dismissal of Challoner every means in my power had been taken to see the property visited frequently by our Representative [Ayers] & all to no purpose – therefore the only thing I could do was to visit the property as often as I could spare the time, which I was sorry to say was very seldom and consequently I was

obliged to obtain from some one on the spot on whom I could rely for full information every week of how matters were progressing & by this means I was enabled to render the concern the valuable assistance before spoken of.

At last, he made it to a full stop. This unpunctuated torrent is reminiscent of his first bitter letter of frustration with Ayers in 1864. He plunged back in, demanding of Beck and Kingston if he had ever clashed with 'any but thoroughly useless officers who they themselves could now see it was in the Company's interests to be rid of?' Neither could answer. With the information he had sought, he had been best able to keep Ayers up to the mark and 'this of course was at the bottom of it all'. This was all part of his feud with the Secretary – and, one senses, more than the Secretary, the man. Rymill kept up his barrage for the best part of an hour, giving his board colleagues some 'good homely truths'. Sir George Kingston's observation was that, in his 30 years' connection with the Burra, he had had no private correspondence on mine matters. New boy Sam Tomkinson responded that perhaps things would have been better if he had. Kingston, the punctilious Speaker of the House of Assembly, seems to have taken a very uncertain lead during his years as Chairman of the Board.

Although Peacock kept quiet, Rymill believed 'he'd have voted with us'. It never came to that; the Furniss resolution was not even seconded. 'Ayers was very warm about it & I was equally so with him.' Rymill went on:

> as I considered that he ought to have been the last to raise any objection. . . . That it is through his neglect of duty that has compelled me to put myself to all this extra trouble – however my duty is simple enough although somewhat unpleasant but so long as I hold a seat at the Board I shall continue to do what I consider is the best for your interest and should only be too glad if Ayers would clear out but this I fear is out of the question.

A mere month later saw the first signs of erosion of confidence in Sanders. A dividend was not coming quickly enough and Rymill had to admit to Graham that he, too, had failed to go on an inspection as promised. In fact, he was one of three directors required to stand for re-election through a poor record of board attendance.

Tomkinson proposed that work be stopped again at the Burra. '[S]uicidal' and an 'absurd idea', reckoned Rymill, as they had made a profit in the last year. Debt might have driven them back to a closure but not as things were now. Unfortunately, continued Rymill, the fact is that Ayers has so . . . [infected?] Tomkinson with his own disbelief in the Mine that Mr. Tomkinson knowing nothing of the prospects of the Mine is thereby lead [sic] into making an Ass of himself.

The board did not go for it, but the bank manager 'could not keep his counsel' and, within an hour, a hundred Burra shares were on offer – Rymill was positive they had to be Tomkinson's and, worse still, one of the Rymills' clients wanted to buy! Despite Rymill's charging Sanders that he was only to put his plans into action if he was confident of success – a mode of operating for which Rymill had previously criticized Sir Henry – all were costly failures. Rymill was sorry they had had anything to do with Darlington or 'his satellites'. They had all come to Burra 'to do their apprenticeships' on the SAMA's time. Wholesale cost cutting was all he could propose, which the 'Trio' opposed, with Tomkinson malleable because he spent a couple of evenings a week at Ayers' house anyway. It really made him 'heartily sick' and he admitted he would have given the whole thing up but for Graham's large stake in the concern. Several reckoned Ayers had some 'special object', Rymill continued:

> Really it makes one think that he must have some underhand game at work in his head – would allow things to drift as they are doing – & I am sure I have hit him hard enough.[?] to make . . . [?] I told his son the other day that I was surprised that his Father who I hold responsible for the bad management of the mine – leading Kingston and Beck to do just what he thinks proper that he did not give it up & save himself all this unpleasantness of working a dead horse – but of course it had no effect.

Then he came to his suspicion: what the board was really aiming for was a wind-up of the concern but, more particularly, Sir Henry was hoping 'to get a good . . . [?] screw as Liquidator'.

The only bright spot for Rymill was the arrival of Harry Graham in the February of 1874. His uncles were overjoyed at the opportunity to entertain him and acquaint him with his colonial birthright; if he were not interested in it, and sold off his assets, it would be a lucrative agency gone for them. Rymill took the young heir to see the famous Burra but it had not left him with a 'very favorable impression of this once thriving place'. Frank and Fred Ayers – good friends of the Graham boys in Europe in the 1860s – were now tarred with the same brush as their father and there was little or no contact between them and the visiting Harry. Similarly, when young Malcolm Graham arrived four years later, there was apparently *no* contact. Indeed, Henry Rymill tried to keep Malcolm away from his own family, being aghast to find him disfigured with cleft lip and palate. Rymill was relieved Malcolm spent most of his time away from town, on sheep stations – very enjoyably – thus saving his uncle acute embarrassment.

With Sam Tomkinson in London, Henry Rymill looked for a sympathizer to join the SAMA board to help balance the influence of Sirs Henry and George. Other influential personalities taking an interest were Thomas Elder and his brother-in-law, the wealthy pastoralist Robert Barr Smith. Rymill had cause to regret, as Sir Henry once had, J.B. Graham's openness over affairs. What Graham had told Barr Smith might cost the magnate a higher price if the two of them and Elder were to buy the mine. The price was now down in the low £60,000s but, given a lift in the world copper price and an improvement in production, the board suddenly felt less keen on selling. Rymill at least thought he finally had an ally at the board table, in William Morgan. His hopes were swiftly dashed, however, when Morgan became the Boucaut ministry's Chief Secretary. Rymill anticipated Morgan's being no more attentive to the SAMA's interests than he considered Sir Henry when similarly occupied.

After years of hoping, a decent find in the Burra raised spirits in 1875, and Sanders went scouting for new ideas around other South Australian mines, though costs had outstripped returns once more. Sir Henry was in the market for shares, which was certainly good for confidence, while Rymill had a proposal for a thorough audit. His choice for auditor was Ayers' competitor for the Secretaryship at the SAMA's foundation, Frederic Wicksteed. Despite Wicksteed's having always been the bridesmaid and never the bride in mining ventures, his audit did not discover anything significant amiss that might have embarrassed Sir Henry.

An Active Backbencher

Although on the backbench, Sir Henry was more active than ever. He proposed a home for inebriates, treating their condition as an illness, not a criminal matter. Was he remembering 'poor Sam Stocks'? Questioning and probing, particularly on financial matters, he always sought economies. Yet, one of his proposals was expensive: no less than the creation of a new ministry – of Agriculture. He was looking for the application of science and innovation generally. Reducing and possibly equalizing salaries to pay for any new ministry was unacceptable to Sir Henry, who felt the Chief Secretary and Commissioner of Public Works bore the brunt. An alternative raised was to replace salaries with mere honoraria. Better still, do it for nothing, Sir Henry rejoined, and it might make governments more permanent! Maybe this comment gave rise to his later reputation as opposed to ministerial pay, yet the laughter he raised makes it clear it was in jest.

The Northern Territory problems festered on into 1875, with recruitment attempted in the Dutch colonies and Mauritius, and ideas about importing 'coolies' from India and China. In the colony, everyone was promised a railway line, with Boucaut and Morgan requiring to borrow £3 million, and very much opposed in this by Sir Henry. This particular occasion is probably the root of the idea that he was completely against government borrowing. Their proposals for backing the loan involved succession and stamp duties, poor Thomas Elder being mocked for naively believing Morgan intended these only for a year. Sir Henry had done the sums and knew these taxes would be inadequate. Morgan believed Sir Henry opposed the borrowing because £3 million coming into the colony would reduce the rates Sir Henry could get on his own and his clients' investments. Sir Henry responded scornfully, and it was a hint of future trouble.

At least Boucaut's government managed to pass a public education bill, which Blyth had failed to do despite a powerful supporting speech from Sir Henry. The secularist Ayers had even been prepared to see Bible readings allowed in schools, without denominationalist commentary, if that would have allowed the bill to pass. Now, Ayers was delighted, although he could not really see where the new measure differed much from his own bill of two years before.

Meanwhile, he pursued further enquiries into scientific developments in agriculture, canvassing overseas experience widely and, a month later, negotiated passage of the Forest Board Bill through the upper house. Barr Smith's Bundaleer station became the home of Australia's first commercial forestry venture. Also, Philip Santo, Sir Henry's former colleague as Foreman of Works at the Burra, attempted to get an anti-capital punishment bill through, though without success. The latest technology was again receiving Sir Henry's close attention as he chaired a Select Committee into the safety and convenience of the coming craze, tram transport. Before the Select Committee was the tram company's young lawyer, Charles Cameron Kingston. Sir Henry would have known his old friend's son since he was born. The next generation to shape Australia was stepping forward.

Two further areas attracted Ayers' involvement. Australian Mutual Provident (AMP) opened up activities in South Australia and he was invited onto its board. He also extended his own, joining the new University's Council, as Treasurer – done gratis. His interest was not restricted merely to financial matters as he contributed to Council's policy discussions. To him was entrusted the delicate matter of conducting the election of the Chancellor and Vice-Chancellor.

Involving as it did a competition between the Chief Justice and the Bishop, Sir Henry was the only one to see the ballot papers.

Education back Home also featured when his old schoolmaster Tom Slade's son invited a donation for the 120th anniversary of Portsmouth's Beneficial Society school – patroness, Her Majesty. With wealth had come approaches for money; he had tried to resist two from acquaintances while in London. In the case of the school, Sir Henry was pleased to donate £20, specifically towards the necessary upkeep of the girls' section. In passing, he praised old Tom, and it was not long before Tom himself wrote soliciting Sir Henry's possible recommendation of his son's teaching aids in Grammar and Geography.

On the Home Front

New Year's Day, 1875, saw the Ayers taking up residence at their new beachfront summer property Seafield Tower. On the Esplanade at Glenelg, the semi-detached building with its central tower commemorated the landmark on the Fifeshire coast dear to Thomas Elder, who occupied the other half. Lady Anne could host parties either for Sir Henry or their bachelor friend. What was it that Ayers had once said in a letter to Graham, that 'Marine Villas' had no charms for him, as only Burra shares held any fascination? Times had changed and with them his priorities. Shares in 'Miss Betsey Burra' were of small value, compared with that earlier euphoric era, and now Sir Henry's interests were much more diversified. He was a wealthy man and could well afford a little cooling sea breeze to blow through his portals at the height of those South Australian summers, which he had found progressively harder to bear.

Meanwhile, Ernest Ayers and Maggie's fiancé, Arthur Lungley, were dodging summer showers as they turned out for the South Australian Cricket Association's XI for a New Year match against the Yorke Peninsula XI. The game did not promise, for Lungley, the same glory that he had enjoyed the previous season when he had claimed the scalp of cricketing giant W.G. Grace, caught for only six off a Lungley ball. As a mere engineer in the Water Department, Lungley was not Sir Henry's idea of a husband for his beloved Maggie. Though his suit was supported by the family through a six-month war of attrition, it took several bruising encounters in Sir Henry's study before Lungley succeeded. Perhaps the young man's Captain's speech at the post-match dinner, chaired by Sir Henry, during the Grace tour served to elevate his chances. Success assured, he apologized profusely for what had been said in the heat of the moment, promising to work hard to prove worthy of Sir Henry's trust. Thinking of Maggie and his grandchildren's futures, Sir Henry made extremely generous provision in setting up a trust fund for them of an initial £18,000. The wedding was 'a very grand one', according to Rymill: 'many hundreds could not find standing room even in the Church – and the Breakfast was something tip-top – forty sat down including the Governor'. A chilly month at Victor Harbor as a suitable honeymoon was easily mocked, though. Rymill knew Harry Graham would concur.

With the wedding came the funerals: Sir Henry's old friend, and Harry Ayers' grandfather-in-law, Sir James Hurtle Fisher, had passed away and the wreck of the *Gothenburg* put a stop to any further revivals of the 'Tommy and Harry' show. 'Tea-pot' Tommy Reynolds was lost aboard the vessel inexplicably coming under both full steam and full sail, in thick weather, through the Flinders Passage. Reynolds' ambitions in the Northern Territory had proved fatal. The shock of the loss of life for the colony was a dreadful echo of the wreck of the *Admella*.

18

CAREER AND REPUTATION IN TATTERS?

A new parliamentary year, in May 1876, saw J.P. Boucaut move his Attorney-General, Samuel Way, to the Supreme Court as successor to the late Chief Justice Hanson. Fred Ayers had already left the Way partnership the previous year to join his brother, Frank, left high and dry when partner R.I. Stow succeeded another victim of the *Gothenburg* disaster, Justice Wearing. Unsurprisingly, Ayers & Ayers became the SAMA's lawyers.

Within days of Way's elevation, Sir Henry found himself back in government when the Boucaut/Morgan ministry fell. John Colton led in the House of Assembly, taking Sir Henry as Chief Secretary in the upper house. The country was amazed, Rymill told an ailing Graham, to find Sir Henry adopting Morgan's 'wholesale railways' policy – what one member described as 'branches on the moon' – which he had previously strongly opposed. According to Sir Henry, however, this was by no means the case; he had never opposed it, only Morgan's plans for paying for it. Sir Henry would firstly assess the scale of the projects, which would reveal their actual cost. From that he could gauge the borrowing requirement. In the event, it would be those probate and succession duties which were largely to fund the scheme. At long last a line would snake northward into the interior from Port Augusta, and an extension was to begin from Kapunda to the North-West Bend of the River Murray – a short cut to the Port for riverine exports.

Also mooted was a bright idea for city rail branches for efficient delivery directly to shops and warehouses, during limited hours, but this was too innovative. A contentious idea for a tramway along the sea-front from Glenelg also came up for parliamentary consideration; it might have passed the Elder and Ayers Seafield Tower. In the event, the sharp turn onto the seafront by the Glenelg jetty was considered too dangerous, especially in the summer season, particularly if new 'silent' motors powered the trams. Luckily, discussion of the delicate issue was conducted on a day when Sir Henry was absent from the committee dealing with it. Instead, it turned out that it would be Judge Gwynne's property which might be affected, and the idea was challenged by his lawyer son, E.C. Gwynne Jnr. On a day when the seasoned Sir Henry was present, he gave whippersnapper C.C. Kingston, appearing for the company, a little advice: on the very practical question of stipulating that the company would have to remake the road surface damaged as the lines were laid down. Kingston thought that was 'surplusage'. 'No, it would not be surplusage', rejoined Sir Henry firmly. If that sort of thing were not spelled out, it could end up in the courts. C.C. Kingston is known to have been a volatile character – he was famed, some years later when a parliamentarian himself, for being apprehended by police in Victoria Square, preparing to meet a fellow parliamentarian in a duel – and his temper may well have been rising during this cross-examination. At one point, with the younger Kingston showing some frustration with Sir Henry's close scrutiny, the knight remarked to him: 'I am pointing this out for the benefit of your clients.' Another legally minded son would also come before Sir Henry's scrutiny when another tram project saw its lawyer before a Select Committee. 'You are Frederic Ayers,' intoned his father, in the Chair.

In the thick of it once more, the Ayers moved back from the seaside to a much enlarged North Terrace house. Two more granddaughters made their appearance, with Maggie's first child, Edith, and a fourth, Agnes, for Fred and Evelyn.

Some indisposition saw Sir Henry forced to legislate from a seated position for nearly a fortnight from late September 1876 – excused by the Legislative Council for needing to do so – during a period in which he carried a bill recognizing trade unions. The 27 year old who had stood in front of the Burra Hotel and refused to negotiate with striking miners en masse now, older and wiser, legislated a legal framework for unions. Speaking in support, and seconded by Philip Santo, he observed that workers should have no lesser rights to combine than had employers. Sir Henry's views on the matter of workers' movements must have crystallized since his time in London, from where he had written his sons of a political discussion at which the topic had arisen. A Swiss cleric, the Reverend Petavel – son-in-law of the late Dr Kent – had spoken of the perplexity of all European governments on the question of working-class organizations. They could not tell what the aims of these groupings might be because of their meetings being held in secret. It was better that unions be formed out in the open, under law, than that they be secret and open to abuse, thought Sir Henry.

A bill for deep drainage of the 'empoisoned' ground on which Adelaide now stood was also brought forward. A commission enquired into the subject, considering 'recycling' and other 'environmental' issues. At long last, the company town, Burra, would also be cleaned up. The SAMA had rarely shown much enthusiasm for a Burra council. Now, the company gave its blessing and Sir Henry tabled the Council by-laws in parliament. Tyneside, the Black Country and Petavel's comments had made their marks on Sir Henry, who perhaps had revised his attitudes to denizens of such industrial wastes.

In October, Sir Henry appears back on his feet again, in good time to return the compliment that William Milne had previously paid him, now that the President had also been knighted.

Sir Henry deployed his most patient diplomacy for negotiations with the Victorian government over the wayward border line between the two colonies. With the population growing through natural increase and further immigration, useful land was in short supply. South Australia sought to reclaim 700 square miles of its statutory territory, demarcation having been hastily and erroneously done under Governor Robe in 1847. The two colonies had agreed their astronomers should look at it but, at the time, Charles Todd – who had done a stint at the Greenwich Observatory – had been busy building the overland telegraph. Blyth had pursued it further with the Privy Council in London as, by 1873, Victoria was quite happy to let matters stay as they were. Once in Sir Henry's lap, he nudged and cajoled for a whole year, his tone painstaking and persuasive, even through a change of Victorian government. Patience was only tested at the last when the reply of his Victorian counterpart was published in the Melbourne press before it reached Sir Henry. This 'divested it of much of its interest', Sir Henry replied cuttingly. Resolution of the dispute was not to come under Sir Henry and it continued to rumble on, causing problems into modern times, eventually needing deflection of the boundary, over the Southern Ocean, when the search was on for offshore oil.

Queensland had its own issue with South Australia as it was apprehensive about any increase of Chinese in the Northern Territory, seen as something of a panacea by the southern colony for its Territory problems. For all the colonies, there were ongoing shortcomings regarding the ocean Mail service and Sir Henry circularized his intercolonial colleagues on the need for vast improvement. The growth in the Australian and New Zealand populations and their financial clout, he maintained, made it imperative. He wanted them to be at no disadvantage compared with India and China. Security of communications was another area of concern to him; there should be another direct cable, to Fremantle, for fear of accidental or deliberate damage to the existing telegraph line.

Not all his intercolonial tasks involved problem solving, however, as he responded favourably to the panel preparing for the Paris Exhibition of 1878. The Prince of Wales' name had been heavily dangled as Chairman of the panel, lest there be any colonial backsliders. As the Burra started to struggle once again, Sir Henry wrote to Captain Sanders requesting suitable samples for Paris.

Sydney was once more the venue for a conference of the colonial heads in early 1877. Not only was duplication of the cable on the agenda, but new links with Singapore and the United States as well. Sir Henry took a circuitous route there by way of South Australia's South-East. His were the scissors poised at the ribbon for the opening ceremony of the Kingston-to-Naracoorte railway. He looked forward, he told the locals, to possible free trade with nearby Victoria. By way of its Western District, he picked up the railway for Melbourne and his Sydney ship. Mid-February saw him back home. Well-received at further formal openings – of Burra's hospital and school – Sir Henry raised doubts about federation. Although he had previously spoken heartily in its favour, he now acknowledged that it would be difficult to accomplish. As a precursor, he could foresee a union between his own colony and New South Wales. Full federation he did not expect to see in his lifetime.

Familiar Faces Depart

News of deaths greeted his return. Old sparring-partner G.S. Walters was gone. So, too, was the former friend who had been central to his business world of 30 years – J.B. Graham. Louisa Maude had not only nursed her dying father but had handled his business matters, with her brother Harry Graham touring the world again. Henry Rymill complimented his niece on her style as, once, a young Ayers had complimented her mother's. The Rymill brothers were put out, however, to find they were not mentioned in Graham's will. Henry Rymill requested, indeed, virtually demanded, that the business affairs of Graham's underage son, Malcolm, be immediately passed over to his Adelaide uncles.

Governor Musgrave was moving on. When he had arrived, in 1873, Sir Henry had stolen a march on his Cabinet colleagues, hitching a ride out into the Gulf aboard the press boat to greet him. Musgrave was the only unknighted Governor since Robe in the 1840s, so Sir Henry had him at a social disadvantage for some of his tenure. His successor, W.W. Cairns, also looked up to the experienced Sir Henry, seeking advice on matters about which he felt uncertain. Cairns stayed only briefly, citing health concerns. Departing, too, was Frank Ayers, after nine years back in the colony. He took with him his sister Josey, touring the United States and Europe en route for Home.

Terminal Realities

For Sir Henry, 1877 was a minefield. The Burra 'Monster' was in its death-throes. Poorer quality produce coincided with a still lower world price for copper and, with Sir Henry back from Sydney, an on-site inspection led to some painful decisions, especially for the town's population. Pettifogging economies accompanied continued exploration at depth but options were narrowing. Independent specialists inspected but gave no hope. The mine closed on its thirty-second anniversary, in late September, with Sir Henry fighting his fiercest political battle.

The initial irritant was the appointment by the Ayers/Colton ministry of a fellow-parliamentarian to succeed the late F.S. Dutton as Agent General in London. Arthur Blyth was a long-time SAMA and parliamentary colleague of Sir Henry, being 'Premier' a couple of times himself, both with and without Sir Henry. Although Blyth had undoubtedly strong credentials for the London position, it appeared a nakedly political appointment.

More controversial still was the fact that work on a new parliament building, next door, had begun just on the vote of the House of Assembly. The upper house, seething at this affront to its Privileges, rounded on Sir Henry who, it felt, had not defended these. In fact, Sir Henry was still aboard ship at the time, returning from the Sydney conference. Citing a Westminster precedent, he ignored a lack-of-confidence motion in the upper house as the ministry still had full support in the Assembly. His opponents smelled blood, however, with William Morgan seeing an opportunity to regain the Chief Secretaryship with its – albeit now reduced – remuneration of £1,000 a year. Morgan's supporter was Richard Chaffey Baker, son of Sir Henry's erstwhile arch-opponent, the late John Baker. Young Baker maintained that the Council *conceded* leadership in the upper house, and therefore could also withhold it. The house withdrew recognition of Sir Henry as the leader, and was incredulous that he would not step down after a vote of no confidence. Where was his self-respect?

In the Legislative Council in September, Sir William Milne gave a President's Opinion which went against Sir Henry, whose position was fast becoming untenable. Still commanding a majority in the lower house, Colton's ministry would not part with Sir Henry, who defended himself powerfully in the Upper. The decision to begin building the new parliament house had been taken in his absence, he pointed out, and Morgan had moved against him before he had even the opportunity to defend the Legislative Council's Privileges. The attack was unconstitutional. To step down would be to admit wrongdoing, of which he was not guilty. He also observed that the attacks had become personal as William Morgan had used the word 'obnoxious'. He might be obnoxious to the Honourable Messrs Morgan and Baker in the same way that they were obnoxious to him – 'Laughter and Hear, hear.' However obnoxious, they would have to hear him for as long as South Australians continued to elect him.

Morgan and Baker, however, would have his political head, as he watched his support dwindle to one or two long-time adherents. With the crisis at its peak, and the press calling on him to end it by falling on his sword, in steamed the new Governor, the elaborately bewhiskered Major-General Sir William Jervois, a defences specialist. His particular expertise was highly valued at that moment because the Russo-Turkish War was raging, bringing the ever-present threat of general European conflagration. Morgan dropped a tome of evidence in the politically inexperienced Governor's lap in the hope of dislodging opponent Ayers. Jervois skilfully sidestepped the paperwork, instead making some generalized pieties – everyone pulling together for the common good.

However, the ministry eventually had to go, Sir Henry accomplishing his move with a good deal of dignity intact, while Colton's performance better suited a 'Chief Jester', remarked the *Register*. While a cartoonist in *The Lantern* had Sir Henry as the target for shell and torpedo practice for HMS *Legislative Council*, his long-time intellectual ally in the Unitarian community, John Howard Clark, editorialized bitterly in the *Register* throughout this period. His friend's achievements in government could be written down on a very small piece of paper. He damned him as a 'mere red-tapist' in any of his ministries, influenced only by 'the most narrow, and self-interested views'. Clark, dying of tuberculosis, was clearly embittered to see ideals they held in common unable to be fulfilled in political reality. Sir Henry's normal approach to political defeats – in Governor Fergusson's judgement – had been to 'take all changes very philosophically', but Clark's public criticisms, in tandem with the unconstitutional nature of this defeat, came especially hard for the deposed Sir Henry. At Clark's funeral, a few months later, he was notably absent, with no apology recorded.

As if Sir Henry did not have enough on his mind, at the height of the political crisis the matter of capital punishment for a convicted murderer came up. Clemency decisions lay in the hands of the Governor, guided by the Cabinet and Chief Justice. From the report on capital punishment compiled for Sir Henry earlier in 1877, it is clear that no executions had occurred during any period when Sir Henry had been in government. This is not to say that he had not sat in sessions of the Executive Council which had actually commuted death sentences – as the Deputy Sheriff had said, there was no list extant of grants of clemency, but an estimate of perhaps ten cases had been made. Until this point, then, Sir Henry had no one's death upon his conscience. In July, Sir Henry and his colleagues had to consider the fate of one Charles Streitman. The unfortunate Dutchman had stabbed an assistant bailiff he had found in his house in Wallaroo. The desperate man, well respected in the community, had been truly penitent, and there had been strong public feeling that his sentence should be commuted. There was some sense that the Executive Council was trying to demonstrate that it would not be swayed by public opinion in such matters, with the decision going against Streitman, who faced his end bravely. One would hope that the consideration of clemency was not decided on hurriedly, in the full heat of the political crisis, with the members of the Colton Cabinet on short fuses. One wonders how Sir Henry slept on the night of the twenty-third, knowing what would happen the following morning.

When his time came, Streitman appears to have been the calmest one present. The hangman had some problem getting the hood over the condemned man's head, rushed to pull the lever, forgetting to pinion the man's legs. The rope-length was wrong and, on his rebounding from the drop, Streitman managed to get his feet on the scaffold again. They were forced through the hole a second time and the unfortunate man hung suspended, his chest heaving for a full 20 to 25 minutes before the medical men present could pronounce him dead. The *Register* expressed the fervent hope that, if a death sentence were to be carried out again, it would be more mercifully efficient in its execution. The protracted end had a very disturbing effect upon the witnesses. Its report must also have had an extraordinarily disturbing effect upon the man's widow and three children. Whether or not Sir Henry got to hear of the way the final scene in Streitman's drama was handled by reading it in the paper or by word of mouth we do not know. If he did not get it first-hand from the Sheriff, W.R. Boothby – who read the condemned man his death warrant – he would doubtless have got it second-hand from Boothby's brother, Josiah. The latter was Civil Service head of the Chief Secretary's department. If sleep had been fitful for Sir Henry on the twenty-third, one wonders even more how he slept on the night of the twenty-fourth after learning how the execution was botched.

Ayers Bides His Time

In the aftermath of the ministry's resignation, Sir Henry was not for keeping his head down. He turned up, bold as brass, to see Governor Jervois inexpertly bend the pin he was driving in to open the first tramline. He joined in the speech making unabashed, and on a constitutional point. He spoke in reply to Henry Scott, the Mayor, who, despite being a shareholder in the tramway enterprise, had seen it as his public duty to fight against the bill setting up the project in his role as a parliamentarian. Sir Henry hoped they would all similarly not seek to advance their private interests against those of the public. The public also had the power to make the upper house more universal if they wished, by further relaxing the property qualifications.

The colonists slept more safely in their beds wrapped in the cotton wool of their defence-expert Governor. There was a chance of Britain's entering the Russo-Turkish war, driven on by the 'jingo'-song, and military man Jervois came up with a naval panacea: if the colony could afford it, an iron-clad vessel with heavy guns should be based at Kangaroo Island, protecting both South Australian gulfs and the approach from the south-east.

The pump-boilers at the Burra were left to go cold and Henry Rymill put the best face on the situation he could for the sake of the younger Grahams, heirs to their father's wealth, including Burra shares. He now wrote monthly to Harry Graham and reconciled the Grahams to the closure: the deepening had only aimed at the 100-fathom level whereas it really should have been to the 120. The shareholders might have had to face a 'call' on their own funds if losses had been left to go any longer, so now the 'widows and orphans' would be spared. As economies, Darlington's retainer was reduced, despite Sir Henry's protestations, and Rymill's placeman in the mine office, Furniss, was retrenched in favour of a cheaper subordinate, William West, who managed the site. Sir Henry and the board went to Burra to see what could be had for the mine's horses when the younger Henderson, son of the Cheshire 'Acid Man', arrived in town looking for a bargain. With their 'mouthes wide open', a figure of £100,000 was asked, drawing no response. Even Sir Henry's salary was lopped to £250, a figure he had not been on for 30 years.

Jervois went Home, briefly, to make his defence report and fetch his family. At a London banquet on the occasion, Harry Graham found himself uncomfortably at 'the same mahogany' as Frank and Harry Ayers. Harry, with Ada and their children, had made his first British visit, resulting in a further Ayers grandson, Harry Cecil, being British-born. Back in Adelaide, just before parliament resumed, in May 1878, Ernest Ayers married Sir William Milne's daughter, Barbara.

Out of office, with nothing happening at the mine, and with a sizeable chunk of the family away, no doubt Sir Henry was glad to be back in parliamentary harness as he was also at odds with his fellow board members of SAGasCo. Whereas his instinct was to compete openly with the Provincial Gas Company, the board wanted to take it over. Defeated at the AGM, he walked out, allegedly stuffing a bag of candles in his pocket on the way home – a nice jibe from the *Portonian* regarding his forfeiture of a free gas supply at North Terrace.

Given the earlier fall from power, the knight's legislating enthusiasm appeared to have cooled and he was pointedly kept off the Select Committee looking into, not only the renewal of work on the new parliament building but also its actual siting. Alternative sites, such as Victoria Square or even on the sacred ground occupied by Government House were considered, though Speaker Kingston jumped to the latter's defence; Colonel Light had always intended the finest position in Adelaide for the representative of the Crown. Governor Jervois had his own sketches for a summer residence – Marble Hill – atop the Mount Lofty ranges and, having struck up a friendship with Sir Henry, had him drop in to Government House to look them over. Despite Sir Henry's dismissal from office, Jervois, as with Cairns before him, valued his long experience of the colony's economic and political life.

Newly knighted Thomas Elder retired from parliament, along with a couple of other experienced hands, including a Sir Henry supporter who liked to write to the papers as 'Saunders McTavish'. Generally, for Sir Henry, the Session was one round of chairing Select Committees.

In Britain, Josey married John Bagot. Her trip there with brother Frank, joined afterwards by Harry and Ada, looks very like a gathering of the clan Ayers for just such a family event. The bridegroom, grandson of Captain Bagot, also sailed to Britain specially for the ceremony, held at Sir Henry's brother John's church in Hastings. Whether he and Lady Ayers had ever harboured any intentions of heading once more for Britain – and he had earlier maintained that he would only go once – nothing came of it. They missed the wedding of their youngest child. The political situation was too knife-edged, with Sir Henry lying in wait for any loss of footing by Chief Secretary Morgan. The younger generation swept back into South Australia early in 1879.

That year also saw Sir William Milne take his overseas trip and, this time, in his absence, Sir Henry could not dodge nomination to the Chair of the Legislative Council. Morgan was safe for the time being, having recently become 'Premier' when Boucaut succeeded the late R.I. Stow on the Supreme Court bench.

With Sir William Milne's return, 1880 saw Sir Henry once more on the floor of the chamber, like a creature waking from hibernation. As only a backbencher, though, he was in no position to influence the government's order of business and so another try at a Married Women's Property Act was elbowed aside in favour of a bill for a pension for the ailing Judge Gwynne. A gender divide was sharply evident, too, over pubs' licensing hours, a petition of many thousands of men outweighing one of only a few hundred 'wives, mothers, and daughters' who wanted an end to Sunday opening.

With the gathering years, more friends and associates departed. At 92, Captain Bagot breathed no more and 'Mr Billy' Sanders did not survive running for the Glenelg train at 79. As 1880 wore on, it became clear that Sir Henry's associate of 35 years – a 'trump' as he had called him at the time of the miners' strike – Sir George Kingston, would be seen no more in the SAMA chair, nor in that of the Speaker. His last speech in the House of Assembly had been to defend Colonel Light's 'gift' to the city of Adelaide, its encircling Parklands, against encroachment. He died on board ship bound for India, with his daughter, seeking some sort of relief for his terminal condition. Referring to 'Paddy' Kingston's period as deputy to Colonel Light, the *Advertiser* closed its notice of Sir George's passing with: 'He roamed about these plains when they were a beautiful wilderness, before the savages and the kangaroo were dispossessed.'

One other name from the past reappeared in the form of a letter from James Coke. He had intended to introduce his nephew coming on from New Zealand although, given the vagaries of the Mails, his letter had taken a year to arrive. Sir Henry sent his old friend a cheque for his namesake; Coke had been fathering children through the 1870s during his late sixties. At 72, he found his work irksome in the Swansea Harbour Board. By contrast, Sir Henry told James he was doing only 'a bit of business, a bit of politics and plenty of leisure'. Remembering Coke's two South Australian interests, Sir Henry remarked that, with copper prices low, no mine was doing well and there was no prospect of the Burra being reopened. James' other interest in the colony, in grain and the milling trade, was also struggling according to Sir Henry; in fact, the colony itself was struggling. The North Terrace house was quiet in comparison with earlier years. Though the Ayers had 15 grandchildren, only the unmarried Frank – 'your old comrade' – was living with his parents. It was a long time ago, but those earlier years had recently been in Sir Henry's mind, as he had shown a painting of Coke's house in Adelaide's Norwood to his son, Harry. Sir Henry was now on the verge of 60.

19

A LONELY APOTHEOSIS: RAW YOUNG EMIGRANT BECOME GRAND OLD MAN

New Year 1881 saw an unexpected gesture of reconciliation from old foe William Morgan. He apologized to Sir Henry for hounding him from office three years before, admitting his grab for the Chief Secretaryship had been unconstitutional, as Sir Henry had maintained. A 'political and moral victory', reckoned Sir Henry's one-time Attorney-General, and gold-panning chum, R.B. Andrews. Upper-house members were delighted with the reconciliation, which Andrews thought a bit rich considering most had sided with Morgan. *Adelaide Punch* likened them to two tomcats, tied by the tails and suspended one each side of a fence. It stopped them clawing each other and, at worst, would only result in a couple of palings being knocked out. A poem followed, beginning:

> YE FRIENDS RESTORED
> Sir Henry Ayers of Adelayde,
> A gallant knighte was he;
> But most unseemly was his ire
> When he didde set it free.
>
> A temper had hee that didde brooke
> No words from man nee boye,
> And if one didde hym contradicte
> It didde hym much annoye.

> Will Morgan, hee was rude of tongue,
> And eke of caustic speeche,
> And when hee bolts of satire shot
> They didde the object reach.
>
> Sir Henry didde encounter hym
> In fair and open fight,
> But who was worsted in the lists
> No man can say aright.
>
> But this is true – betwi'xt those too
> There raged a deadly feud,
> And when they met they oft exchanged
> Expressions that were rude.

After several more verses about the ongoing feud, reconciliation followed:

> 'Twas sweet indeed to see they two
> That erst didde each despise,
> Go walking forth in soft commune
> And eke in lovynge guise.

March brought the spectre of family mortality. On the sixteenth, Sir Henry wrote 'Harry's boy, Harry Cecil died.' On the twentieth, their little girl of six, Genevieve, followed. The terrible toll was completed with their Frank lost as well. Sir Henry bracketed the *Memorabilia* entries with one word: 'Diphtheria'. It was a cruel blow for three Ayers grandchildren to be wrenched from the family at a stroke. Attempts at isolation had been made; two were nursed at Seafield, while little Frank succumbed at North Terrace. This perhaps saved baby Sidney. One wonders what sort of company Sir Henry and Lady Anne were, two weeks later, as guests at Alexander Hay's neo-Gothic seaside home. How far could five days' Easter escape go in assuaging the trauma?

Polling day immediately followed. Sir Henry, though placed sixth, was one of only two members re-elected. His success was viewed as a miracle, so sick were the voters of the subject of Legislative Council reform; henceforth an upper-house term was reduced from a standard twelve to nine years.

The new parliament, sitting in June, needed to replace retiring President Sir William Milne. Sir Henry was elected unopposed, proposed by an R.C. Baker no doubt delighted to see him elevated above the ranks of potential Chief Secretaries. Were there any knowing glances at the half-finished foundation-work next to the old parliament building as the worthies passed en route to present their choice to Governor Jervois – who declared himself 'satisfied'? The cause of Sir Henry's last fall from power lay as a rubbly old reminder. A joint committee of both houses would further consider the vexed question of a new Parliament House. Ironically, the only name to go forward automatically was Sir Henry's, as President. He had first sat on such a commission almost 20 years before.

The *Register* editor conceded that, no matter what one felt about him as a politician, he was eminently fitted for Presidency, with long parliamentary experience and an intimate acquaintance with constitutional procedure. His undoubted ability and great influence among members marked him out; with many new members, his experience would be very valuable. He had had a 'larger share in Governmental honours than any other member of either House'. He would be missed from the floor of the House, though, and would be a severe loss as Leader of the Opposition.

Fred Ayers followed his father in a modest involvement in public life, elected to the University Council. Female education – always dear to Sir Henry's heart – received a boost when Royal Letters Patent allowed the first 'lady' to begin a degree course. Sir Henry had experienced some difficulties in investing the £40,000 combined benefactions of Hughes and Elder, with the former's half hedged round with tight strictures. Until the deed could be altered, Hughes' £20,000 had to sit in the bank, at poor interest. With their father turning 60, Fred's brothers Harry and Ernest had already shouldered more responsibility for the private business affairs through the formation of H.L. & A.E. Ayers.

Sir Henry was again in a prominent societal position for another, unexpected, royal visit, by two sons of the Prince of Wales, the Princes Albert Victor and George – the later King George V. These 'boy Princes', as Frank Rymill called them, were mere midshipmen aboard HMS *Bacchante*. Not for them a reception with mayor and councillors bowing and scraping, nor kerfuffles about who would ride in which carriage. In fact, there was a hint of republicanism in the correspondence columns. Tree plantings, school visits and the like were largely the limits of the boys' activities, though they did open the National Art Gallery. Sir Henry dined with them quietly at Government House, and his postman, Chapman, was musician for a Town Hall ball. The young ladies honoured by dancing with their Royal Highnesses were then 'content to pass away!' commented Frank Rymill.

The Adelaide International Exhibition, a hurriedly assembled 'private enterprise' affair, opened in July, in a crush in the new Art Gallery. Sir Henry joined a procession with plenty of 'rank and title' in it. Later, a little flotilla came up the new 'sheet of water' provided by the recreated Torrens Dam, Sir Henry and his parliamentary colleagues steaming along in *Pioneer*. As an old resident, he toasted 'The Council'. He had no objection to paying rates to add to Adelaide's attractiveness, unlike his friend Tomkinson, who liked to point out where the City Council got things 'a little wrong'. This raised a laugh. Sir Henry said he had once walked rough-shod in the very place they had just sailed over. Old colonists like himself were amazed at the advances.

The next *Memorabilia* entry, for 13 August, is stark:

> My dear Wife died   Born 28 Nov. 1813
> Aged 67 years 8 months
> And 16 days

His customary underlining, when noting a death, is clearly not sufficient: he surrounds the opening words with a box. It is quite telling that, from this point on, *Memorabilia* entries become much sparser. He notes Frank Potts' birthday nearby, perhaps fearing, with his help-meet of 41 years gone, that it might slip his mind. Anne herself, of course, was older than he indicated.

He kept a newspaper cutting, himself correcting her time of death to the Saturday night. The obituary notice indicated that she had had some kind of fever and, although Sir Henry had decided to miss a breakfast at Prince Alfred College, no 'fatal danger' had been anticipated. Her loss would bring sorrow to a large circle, believed the writer, for Lady Ayers, an 'Old and esteemed Colonist', had been well-known in Adelaide for 40 years. One of the guides at Ayers House, an ex-nurse, speculated that typhoid carried her off; a doctor friend of hers concurred, given the timing and nature of the symptoms. The recent bitter blow of the loss of three grandchildren perhaps added to her debility.

On the fifteenth, the coffin left North Terrace at 11.30, for West Terrace cemetery. Sir Henry was well-supported by parliamentary and business associates. A.J. Baker joined Lady Ayers' sons, except Ernest, who was overseas. Nephew-by-marriage Henry Rymill was not present, nor was his brother, Frank. On this sad occasion, though, there was no snub involved as Henry and Lucy were visiting Home and Frank may also have been away.

At 60, Sir Henry was a widower. The North Terrace house would be even quieter now. As Sir Henry lay in his lonely bed, he must have reflected, perhaps with gratitude, on the life he and Anne had carved out for themselves after taking the risk of following her brother, Frank, to the end of the earth. His rise in South Australia's business and political world had enabled Anne to enjoy almost a decade of life as a Lady, a status to which she could never have aspired at Home. Indeed, the memorial over the Ayers cemetery plot inexplicably further elevates her as 'Dame Anne Ayers'. The day after the funeral, Sir Henry dutifully took his place in the President's chair, but the Council adjourned for a week in recognition of his loss.

In Sir Henry's scrapbook, next to Lady Ayers' obituary, is that of William Christie Gosse. The namer of 'Ayers Rock' died at 39, failing to recover from a burst blood vessel. Sir Henry and his Sunday-morning gossip, Dr Gosse, were united in grief within a day.

Near these sad memorials, Sir Henry pasted a clipping from the *Cambrian*, notifying of the birth of twin boys to James Coke's wife. Sir Henry's old Swansea friend was still fathering '[his] dear little children', the fifth and sixth of his old age. James, approaching 74, wrote that despite some recent 'internal growling' he 'did not hear the little chap speak'. He was soon to find that the little chap was only half the story. James wrote warmly of 'Mrs. Ayers' who was 'so kind, gentle, and unassuming in her manners'. He closed with the pious, if lugubrious, thought that he was preparing himself for 'that great and momentous change which we all <u>must</u> undergo'.

Sir James Fergusson, writing from Bombay [Mumbai], could doubly empathize. He had lost Lady Fergusson while Governor. His second wife, too, was now gone – he sent her photograph, taken just three days before her death, to illustrate how quickly things change. Cholera had carried her off hours after their baby was born, dead. Her picture would bring back memories: she was Olivia Richman, daughter of Sir Henry's first colonial employer.

The minister of the Unitarian chapel, John Crawford Woods, gave thanks for Lady Ayers' kindness to his wife through *her* long and painful final illness. The fact that there is not a word about God or expectations of eternal life – no religious overtone whatsoever – is quite arresting. It may underline the 'rational' nature of the beliefs of the small but influential Unitarian congregation in which Sir Henry had found a spiritual home.

Before the many letters of condolence could pour into North Terrace, Sir Henry sat down at his writing desk to, tenderly, give Frank Potts the sad news. The new widower may have been unable to think straight as he misdated the note. His grief did not prevent him from breaking the news to his wife's little brother as gently as he could. Perhaps he thought of the occasion on which scruffy Frank – apparently never a dresser, with sleeves invariably rolled up – on announcing to the Ayers' butler that he had come to visit his sister, had been sent round to the back door.

Despite the shock of losing, by now, five children, Harry and Ada bravely enlarged their family again, with the birth of John Morphett Ayers in May 1882.

Although they still had Mary Elizabeth and Amy surviving by 1881, it must have seemed desirable to have a back-up to remaining heir, Sidney Hurtle.

With Anne's death, the societal centre shifted to *Dimora*, Harry and Ada's gracious new home on East Terrace. Now the Ayers matriarch, Ada assumed the role of hostess for evening parties. However, as Josey's son, Walter Hervey Bagot, tells it – he was a small boy during the coming period – Sir Henry did not want for family company. Daughters and daughters-in-law often brought the grandchildren for lunch to North Terrace, alleviating the 'large and echoing' atmosphere of a house so 'full of gaiety in Grandmama's days'. On Sundays, at least two sons and their wives would join Sir Henry for dinner. Housekeeper Bridget Galvin – the 'household deity' – had a soft spot for the little ones, providing them with lemonade, gingerbread and nuts. Walter felt as though his grandfather never actually noticed him, though, and he explored the 'empty old house with some nervousness'. The wine vaults seemed 'of the most ample and spacious description'. He noted the china labels, marked claret, port, or sherry. His grandfather had carefully selected those in London.

Sir Henry must have been reasonably content with how married life was turning out for Maggie and Arthur. A month or two after his mother-in-law died, Arthur Lungley wrote thanking Sir Henry for a gift. The couple had two daughters and, possibly on the strength of Sir Henry's latest gift, a third was conceived. In total, Sir Henry underpinned his daughter's future with £50,000. Lungley was 'very grateful, and even more obliged'. He felt that it showed he had proven himself worthy of the confidence Sir Henry had placed in him. As a more sentimental gift, Sir Henry passed Lady Anne's Prayer Book on to eldest grandchild Mary Elizabeth.

To commemorate Anne, Sir Henry contributed several hundred pounds to the North Adelaide Cottage Homes charity. Seven small cottages for the needy were immediately required. By the July of 1882, Sir Henry's gift in fact provided some nine cottages for deserving cases, built in Kingston Terrace: the Lady Ayers Homes. The grateful charity spoke of its benefactor's support as typically generous and unostentatious.

Life Goes On

Sir Henry keenly watched what was seen as the nineteenth century's march of progress. The imminent provision of electric light to Adelaide threatened Sir Henry's and others' interests in the SAGasCo, which had monopolized the city's lighting for two decades. With no electricity supply in the other Australian colonies, Adelaide would be at the forefront in legislating for this new energy source, which could power anything from sewing machines to torpedoes, a 'clean, silent' electric railway in Berlin and the lighting of Richard D'Oyly-Carte's new Savoy Theatre in London. The cheap option of hanging wires off poles could cause a tangle with existing telegraph, and now telephone, wires. Turning the new threat into an opportunity, Sir Henry's SAGasCo, through agent Fred Ayers, requested the right to also supply electricity.

With modernity for Adelaide still in the offing, Henry Rymill and family returned to the colony. 'Home is so much more in advance of here', Rymill lamented to Harry Graham. Unlike the proud South Australian that Sir Henry had become, one feels Rymill would have left at any moment, if circumstances permitted. One good harvest, he thought, would cause the progress of the 'Village' to become extraordinary again. Whether he was picking up on Sir Charles Dilke's description of Adelaide as a 'farinaceous village', or whether that was simply the way his adopted city now struck him, is an open question.

For those who had not prospered as a Sir Henry had done, the shape of an Old Colonists' Association was thrashed out in a public meeting. Sir Henry joined the committee as Treasurer, finding a new sphere in which to find satisfaction for the rest of his days. Would only those who had arrived before 1856 – Pioneers and Old Colonists – benefit if they fell on hard times? If so, the Association would be designing itself to disappear before too long, like the Waterloo Veterans. Sir Henry favoured a virtually open-ended vision, in which descendants of those early colonists would be eligible for help, as a continuing recognition of the important services of their forebears in establishing South Australia. Sam Tomkinson reckoned it a guarantee of pauperism, 'offering a premium to idleness'. Back came Sir Henry: 'Riches' could 'take unto themselves wings' and fly away very suddenly, come their grandchildren's generation. In any case, he added, there would be no *entitlement* to financial help; cases would be considered on merit. Sir Henry moved acceptance of the regulations, seconded by Chief Secretary J.C. Bray, the first South Australian-born holder of that office.

Having accepted the umpire's chair in the Legislative Council, there was no going back into active politics. The prestige of his position meant that he would inevitably be called upon to chair this, open that, or generally be present at ceremonial or worthy occasions. He might judge, for example, a debate between members of a Literary Society, on such topics as the desirability of federating the whole British Empire. These involvements, though, were mundane compared with former demands upon his energies and intellect. A *Register* parliamentary sketch writer, in 1885, wrote approvingly of him in his Presidential role, as the member best fitted for the responsibility; he was as 'firm as a General, and discreet as a match-making mother'. He was a bit crotchety, but then so were all men of his age.

At 65, and with a solid 12-year record as University Treasurer, Sir Henry found the post no longer manageable. Given the university's growth, he recommended a committee should handle its finances. Chancellor and Vice-Chancellor gave a warm vote of thanks for his long and valuable service, and he saw out his term as a Council member whose particular interests were, unsurprisingly, in the Building and Finance committees. In unveiling the grandchildren's plaque to Sir Henry at the university, in 1910, Chancellor Way underlined his integrity and the fact that he freed the institution from 'a heavy incubus of debt'.

That year – 1886 – saw the Golden Jubilee of the Province's founding and, while Sir Henry's experience did not stretch back quite to its beginning, his years and position there singled him out for a leading part in celebrations. With Sir Henry in the Chair, on 27 December, 1,300 turned out to an entertainment in the Town Hall. A tableau represented Inspector Tolmer's Gold Escort of 1852 – with Tolmer playing himself. Another represented McDouall Stuart's crossing of the continent, the proceedings interspersed with patriotic songs and fantasias on the mighty organ. In his address, Sir Henry mentioned the likes of Captain Sturt and Colonel Light, the main burden being that 'much had been achieved'. The colony had local offshoots of all the benevolent institutions back Home, but he considered it a pity steps had not been taken, years before, to make their presence unnecessary. Maybe he was ruefully reflecting on things left undone during his tenures of the top office. Nevertheless, it was a 'land flowing with milk and honey' through industry and sacrifice. The future would be as bright as the past, if mistakes such as those of the previous few years could be avoided. His audience greeted that hope with wild enthusiasm.

His reference to recent mistakes, coupled with his earlier comment about riches taking flight, was apt. In the February, the Commercial Bank of South Australia, less than a decade old, had been forced to close its doors. In London it barely caused a ripple; in Adelaide its impact was an 'unparalleled catastrophe', considered the *Register*. Foolish lending practices of the Commercial's Manager – ominously named Crooks – were not his only sins, for he and his Accountant, Wilson, were strongly suspected of embezzling large sums. Chairman of the Board R.A. Tarlton shakily faced the angry and excitable shareholders in the Town Hall. He was loath to attempt to chair the meeting, particularly after witnessing Crooks' entrance to hisses, and calls to send him to the Stockade, or worse, that he be lynched. Tarlton desperately looked for Sir Henry's arrival. As a founder and co-trustee with Robert Barr Smith, he made his entrance and was voted to the chair. He immediately set about getting those assembled on side: everyone regretted the reason for the meeting but, in the interests of themselves and the community, he asked them to enter into the subject 'calmly, dispassionately, and with [their] best judgement'. He was prepared to accept their indignation but asked that they listen carefully to the statements to be made. He proposed a committee of shareholders to investigate the situation. He astutely invited the throng to assist him in keeping order.

Managing two such large and volatile gatherings seems to have been Sir Henry's contribution to saving the situation. Barr Smith's task was to attempt actual resuscitation through London, though to no avail. Crooks, prosecuted for embezzlement, confessed all, owning his shame and exonerating the credulous directors and Accountant Wilson, who received a relatively light sentence. Any halo once glowing above Crooks for catching out W.G. Grace, off Arthur Lungley's ball, had drastically slipped. He would not plead for leniency and Chief Justice Way saw he got none, sending him down for eight years with hard labour.

It was a dreadful period for South Australia, generally, with drought, depression and extensive unemployment. The devout colony considered a day of Humiliation and Prayer advisable. Even that rock-solid institution, the Savings Bank, on whose board Sir Henry had sat for 30 years – chairing for numbers of them – faced a run; its manager rushed into print, steadying nerves. The *Register* editor looked forward, not to a repeat of recent years of excess, but to a less exciting period of steady, sober progress.

The Jubilee celebrations enabled those present to forget for the moment any painful losses of earlier in the year. The Town Hall now furnished the backdrop for the roseate nostalgia for days and personalities long gone. Frank Potts had of course sailed aboard the *Buffalo* but, despite the offer of a complimentary ticket, was not present there nor at the ceremony at the founding spot, Glenelg's Old Gum Tree. Sir Henry felt that the efforts of these pioneers were in danger of losing respect, while the *Advertiser* felt the whole Jubilee celebration would help revive a sense of community, slipping away in days of democratic individualism. Relics were on display, such as Colonel Light's sword. Pictures of Burra were loaned by Sir Henry, and it was suggested that these might form the nucleus of a national collection.

A further celebration the following year, at the Adelaide International Exhibition, coincided with the Queen's Golden Jubilee. Most unusually, the Ayers name does not figure at all in its organization, although, as President of the Legislative Council, Sir Henry appeared in the opening parade. On the shortest day of the year, it drizzled, with thunder and lightning outshining the celebratory gas jets. This disappointed South Australians, anxious to put on their best show for distinguished visitors, including the Victorian colony's Alfred Deakin, who would one day go on to become Australia's Prime

Minister. Sir Henry, on the dais with the notables, was uncharacteristically mute. Charles Todd, indispensable at such events, bowed low before the Governor before transmitting the colony's congratulations to the Queen. Sir Henry assisted Henry Mildred's son, Hiram – all three, Portsmouthians – in pulling together another exhibition of the colony's relics. It was an unusually low-key involvement.

If there were no great excitements left in the public sphere for Sir Henry, were there any remaining in connection with the SAMA, his original arena? There was little happening at the Burra. The odd very modest dividend came largely out of the sale of assets. Ayers saw board associates come and go. Former foe Morgan was Chair in 1882 but, following embarrassing losses in a New Caledonian copper project, he returned to Britain, with his own KCMG. He had little time to enjoy it; he was dead before year's end. Sam Tomkinson came and went, a bit of a will-o'-the-wisp. Ernest Ayers joined the Board and Frank Rymill, who had succeeded his brother, became Chair.

Optimism suddenly rose in the Burra community during 1887 with news that the mine was up for sale. The directors conveyed the mine to an F.J. King, who declined to complete as the copper price plunged. Action against King was started by July 1889. Sir Henry managed all this by telegraphy with some degree of privacy and security, using the code-name 'Orpheus' for their London agent, Morrison. He had earlier dubbed Darlington 'Kooringa' as his *nom de télégraphe*. If the choice of Morrison's alias was Sir Henry's, it would reflect his earlier characterization of all the personalities involved in the negotiations with the PCC in terms of Greek mythology. Perhaps Sir Henry now saw Orpheus charming the beasts prowling around the stricken SAMA. Maybe 'Betsey' had become Eurydice.

Governors' Confidant

Sir Henry found himself in position for a little gloat, with the opening of the first half of New Parliament House – a more comfortable new home for the House of Assembly – on the very site that Sir Henry had suffered for in 1877. He and Speaker Bray hosted an 'At Home' in honour of the occasion. The sounds of the Police Band wafted into the chamber from outside the gallery in which the reporters were 'doomed', said the *Register*, to pass entire Sessions. Governor the Earl of Kintore and his Countess were welcomed by President and Speaker, as they led the dignitaries into a 'Legislative Palace' undreamed of by those who could remember Government House when it was a Hut. Marble and granite were liberally splashed around and, when there were sufficient funds, it was hoped that a balancing Legislative Council chamber would fill up the still empty plot next to King William Street; the wait would be another 50 years. Sir Henry continued his gloat by hosting a dinner in his North Terrace home for Kintore, Bray and about 40 MPs.

Having an actual peer as Governor was a novelty for South Australians. Whereas it is said to be one of the strengths of monarchy that a long-serving monarch such as Queen Victoria – indeed Elizabeth II – has seen many a Prime Minister come and go and can advise from a depth of experience, in South Australian terms, the boot was on Ayers' foot. His length of public involvement was unmatched and so it was his 'advice' that was sought by Governors on all sorts of matters.

Discretion, it appears, was something all Governors could count on in Sir Henry. When Sir William Robinson was nonplussed by a letter in the press from 'A WIFE AND A MOTHER' of North Adelaide, Sir Henry was the man to consult. The correspondent was annoyed that Lady Robinson's absence of a couple of years had meant no entertaining had gone on at Government House, so that not enough of the Governor's salary was being spent among local tradespeople. Possibly she was more aggrieved that her daughter was being deprived of an opportunity for débutance. Sir Henry had a quiet word with the newspaper's editor and, the following day, the paper carried a list of entertaining done in the big House, showing the complainant to be plain wrong. Possibly coincidentally, within a fortnight, Governor Robinson sent Sir Henry a note indicating that a Companionship in the ancient Order of the Bath was imminent, but it never arrived. If Sir Henry's tongue hung out for further such recognition, it would be dangling there for another seven years. Once he had retired from public life, he was awarded the Grand Cross of his Order of St Michael and St George.

Invitations to and from Sir Henry and Lady Ayers had travelled up and down North Terrace over the years. Sometimes Lady Jervois, or her daughter, might request Anne Ayers to assist them in providing Christmas dinner for the invalids of the hospital and the Destitute Asylum. Various Governors invited the couple to dine at Government House, when such guests as Lords Normanby, Carrington and even the Duke of Manchester were visiting. Sometimes, when the invitation had come the other way, a note from a Governor or his Lady might plead a bad cold as real or diplomatic excuse for inability to accept. A very kind invitation to Sir Henry, a month after he lost his wife, came from Lucy, Lady Jervois. Would he like to dine with them? They would be 'quite alone or at most shall ask one other friend'. Clearly, they had quite some thought for him at a very sensitive time. These small parties continued, including at the summer residence, Marble Hill, risen from the plan that Jervois had asked Sir Henry to look over. He could even stay overnight. When it came time for the Jervois to leave the colony, Lucy Jervois requested a photograph of Sir Henry to take with them.

When the Musgraves had moved on, the departure had clearly been quite a break. As a parting gift, Sir Henry offered them a certain picture from his collection. This the Governor was kind enough to turn down, despite Lady Musgrave's unsuccessful attempt at a copy. Over a decade later, Lady Musgrave wrote at length to Sir Henry, from Brisbane, thanking him for his kind letter of condolence on Sir Anthony's death. She was very touched by this letter coming from such a 'highly respected' man, and later requested permission to publish it.

Quiet parties at Government House continued to include Sir Henry during Earl Kintore's time. The inevitable whist could include the late William Peacock's son, Caleb, together with the Brigade Major of the colony's military. On one occasion, in 1891, the invitation was a touch unappetizing: Kintore warned Sir Henry that influenza was 'stalking through the household' but his cheery hope was that 'someone [would] still be alive to welcome [Sir Henry]'.

Such free-and-easy access to the social side of Government House made it convenient for Governors to seek Sir Henry's opinions on more serious matters. A Governor could run an important speech past him, touching on matters in the colony's history which Sir Henry had of course lived. In 1891, Kintore showed him, unofficially, his correspondence on the administration of the Northern Territory. Since Sir Henry had been in on the ground floor of South Australian involvement there, almost 30 years before, Kintore told him he had proposed that a coming federal government should take it on. Earl Kintore was too optimistic about having responsibility for the Northern Territory lifted off South Australia's shoulders by 20 years and, although today it has its own legislature, it is still not a fully fledged State within the Commonwealth.

In 1888, Sir Henry offered himself for the fifth time for the consideration of the voters. He was so apt for the post of President, was the comment, that it would be unthinkable that he would not be returned, which indeed he was.

While he kept busy, his old friend James Coke reached the end of his working life at the age of 81. Sir Henry kept a clipping of the *Cambrian*'s report of the AGM of Swansea Harbour board, noting the retirement of its long-standing Accountant. As not quite such an old dog, Sir Henry must have felt there was still a bit of life left in himself, at a mere 67. The SAMA, the Savings Bank, SAGasCo and the AMP continued to absorb his time and interest. He was always in demand to hold the financial reins of any important testimonial. For several years, he chaired the appeal and monitored the funds for a fitting memorial to Colonel Light. 1891 brought his seventieth birthday and company by-laws of the AMP uncompromisingly required his resignation.

He did not lose his business touch, though. The long-running legal case over the failure of King to complete his purchase of the Burra Mine came to a head in 1893. Darlington telegraphed the question: was he to agree to a compromise to wrap up the case, so long as there were no costs to be borne? The answer was yes and, by the April, the Association was awarded damages of £11,000. The directors, led by Frank Rymill, recorded thanks to Sir Henry, desiring to 'place on record their sense of the zeal, ability and judgement displayed by the Secretary, Sir Henry Ayers, whose arrangement of the facts and evidence, in support of the case, greatly contributed to the favourable result of our suit'.

Calling it a Day

The next logical step would be relinquishing the oversight of debate in the Legislative Council. Sir Henry let it be known during the later months of 1893 that he was considering retirement from politics. As that year's parliamentary Session headed towards its climax, the usual scramble to steer legislation through the upper house put pressure on members, and most particularly 72-year-old Sir Henry. Sittings already lasting until midnight made it necessary to consider calling the Council at 10 a.m. rather than 2 p.m. On the first morning Sitting, on 19 December, the Clerk to the Council read out a message from Sir Henry to say that he would be unavoidably absent. Lance Stirling, the Deputy President, took the chair, also having a brief-and-to-the-point message: Sir Henry begged to tender his resignation, as President and indeed as a Member of the Legislative Council.

A mid-November edition of the *Advertiser* had revealed that he would go in a few weeks. He found the long sittings 'irksome', and his decision about when to go was partly governed by the fact that there was a forthcoming election. It would save the colony – at a time, again, of economic stress – the cost of funding a by-election, should he have to retire through ill health. No parliamentarian in any of Her Majesty's Dominions could match his nearly 37 years in the Legislative Council. Only Gladstone's 60 years in the House of Commons put Sir Henry's service in the shade. He could not be begrudged his retirement, then, and he would have the 'satisfaction of knowing that he had earned, in a truer sense than it is given to most men to earn it, the . . . [?] of "well done," which is the reward [?] of the "good and faithful servant"'.

The *Register* said he had taken an active part in the destiny of South Australia and was a 'Patriarch in Politics'. Some of the legislators had not even been born when Sir Henry had entered parliament, or were more interested at that time in 'lollies, tops, and kites, or wanted a loan of the moon for a plaything'.

The recent irksome Sittings had not only been long but also fairly trying for an elderly President. In earlier November, Ebenezer Ward, hair tousled and in a 'pugilistic' frame of mind, had had a run-in with Sir Henry. He had wanted Government business dealt with before Private business on a Wednesday, which Sir Henry would not allow. Voted to silence by his colleagues, Ward later moved some motion about flag-raising at Government House. This gave rise to another 'exciting exchange' with Sir Henry, until it was moved that Ward be not heard for a second time. It all helped to confirm Sir Henry in his resolve to go through with his intention of calling it a day. Despite the advance signs, no one could really believe that it had happened. One member felt as bad as if the vacancy had come about through death.

Business resumed, then Ebenezer Ward – the Council's 'stormy petrel' – butted in. His Point of Order was that nothing they said and did was valid because they had no President and were ill-constituted. He won his point and the members chose Sir Henry's successor, Richard Chaffey Baker, Leader of the Opposition. Baker complimented his retired predecessor for his 'ability, urbanity, courtesy, and impartiality'. He was the only sitting member to have known Sir Henry active on the floor of the House, had had the honour of proposing him twice as President and felt honoured to be succeeding him. Tomkinson characterized his friend as the 'ablest statesman' in the colony's parliament. Another member, Dr Campbell, said he had been a friend and adviser to all members and it had seemed as though he would never grow old, causing them to lose his familiar face from the chamber. The youngest member, W.A. Robinson, as he said, on the threshold of his political career, had been well advised by Sir Henry, who had been like a father to them all.

What had determined Sir Henry's 'unavoidable' absence from the chamber that morning is open to conjecture. George Goyder, Surveyor-General of the previous 30 years, had just announced his retirement, and that may have galvanized Sir Henry to follow suit. News of the death of old friend William Paxton may also have undermined Sir Henry's zest.

The suggestion of a public event to mark Sir Henry's retirement was one he graciously declined. Denied this possibility, the people of Burra expressed their appreciation of their long association with him, in the form of a written testimonial. They believed he had made a 'mark in history that [would] always be remembered'.

'I am going to write your political epitaph', began the editor of the satirical magazine, *Quiz*. He said he 'normally agree[d] with that cynic, Byron, that epitaphs should no more be believed than women' but, in this case, the writer was 'in earnest. When the history of South Australia is written, say fifty years hence, the name of Ayers will be deservedly prominent' – how wrong a writer could be! In any histories, written contemporaneously or since, Sir Henry's name barely figures. Thirty-six years in parliament, seven times 'Premier' and serving in 11 ministries, Sir Henry 'must have possessed a force of character that cannot possibly be overlooked by the historian'. His 'dignity and impartiality' as President of the Legislative Council was proverbial.

He had seen successes and failures, booms and busts, and yet he could sit in his study and coolly put his finger on shortcomings and deduce 'hopeful signs for the future'. The writer envied Sir Henry his serenity, weighing up pros and cons carefully before deciding. He had thought of Sir Henry as a Liberal, in several of his governments, who had called for parliament to be renewed by the influx of younger members, and this before the entry of Labor members. As President, Sir Henry had become less wedded to 'old world opinions'. The writer went on, 'You detest sham and you admire honesty of purpose, even though that honesty may appear to you to be misdirected.'

Sir Henry had known the 'emoluments of office' through six years, and yet he had also known the 'cold shades of opposition. You were a fair fighter. You did not believe in stabbing in the dark. When you opposed you did so openly.' Sir Henry, though, was not a 'bitter oppositionist', that was to say 'bitter of tongue'. His manner had always been one of courtesy to an opponent. In public speaking, the *Quiz* writer rated Sir Henry the best of South Australian speakers. He did not have the 'flowery ease' of a Lieutenant Governor (Chief Justice Way) who could 'rhapsodise about anything from an abnormally sized turnip to a Salvation Army lass's bonnet'. Nor had he the polish of Mr Symon, QC, nor the 'glittering periods of Mr. Rowland Rees, nor the forceful oratory (eloquence of the taproom) of Ebenezer Ward'. Yet, Sir Henry's style was 'gracious and the matter [was] immediately good'. A reporter was the best judge, claimed the writer, and Sir Henry's words were always simple but suitable; they could be transcribed from notes without alteration. He was never stuck for a word. Sadly, in the previous 12 and a half years – since becoming President – Sir Henry's voice had been heard too seldom. He had become a judge rather than a participant.

The writer wished he would continue to advise younger members. He should write as well, of his experiences in South Australia, and in simple language, not like the efforts of Sir Henry Parkes. Maybe the writer could be Sir Henry's biographer, he speculated. He felt he had seemed to flatter the old man, yet claiming to be a cynic. He recognized Sir Henry as 'an honest man,' a man 'with whom soured old Diogenes would have been delighted to have associated'. When there was nothing left to say, the writer could only wrap it up by stating: 'I believe in you.'

20

DEATH AND RESURRECTION

Retired from public life in his early seventies, Sir Henry rattled round in the large, mostly empty house on North Terrace, listening for the echoes of family life, and the celebrated 'Evening Parties', of the past 40 years. His interest in adding entries to his *Memorabilia* was petering out. A note of his resignation from the Presidency of the Legislative Council provides the only entry for 1893. He notes the award of the Grand Cross of his Order, the following May, balanced out by the domestic-level final entry recording the arrival of one more grandchild, Erlstoun Barbara, a second daughter for Ernest and Barbara, in 1894. She would not survive toddlerhood.

There had been too many unhappy entries of later years. '<u>Dear Maggie died in her 39th year</u>' was the poignant entry of September 1887. Sir William Milne wrote in sympathy, not needing to mention the earlier loss of his own young daughter, Mary. Another old friendly voice came from London: Sir Arthur Blyth said Maggie had been 'one of the most unselfish beings' he had ever known.

Sir Henry kept a poem, capturing the devotion shown by Lucy Rymill's sister, Amy, in nursing their mother, Margaret Baker, through a difficult final illness in 1890. The poet was their bereft father, A.J. Baker. Lucy Rymill, herself, had already died of meningitis – 'or inflammation of the membranes of the brain and spinal marrow', as Sir Henry notes it – on her forty-sixth birthday. Henry Rymill had soon followed his former mentor down the widower's path. Margaret Baker's sister, Elizabeth Churcher, had died 16 years previously. Now, Frank Potts joined his stepsisters. The inveterate worker with his hands, in addition to building sailing-boats, steamers, winery equipment and even a piano, according to his descendants, had fashioned his own coffin. Ever practical, he had used it to store his apple crop pending its ultimate use. Lady Ayers' South Australian siblings were all gone.

Death had not yet done with the Ayers grandchildren, either: 1889 saw the loss of Josey's boy, Charles Hervey Bagot, at three, and in 1891 Harry and Ada lost another son, Lancelot, at seven. In memory of her six lost children, Ada donated a beautiful, opalescent, Tiffany-designed window to St Paul's church, Pulteney Street, in 1909. One of a pair, the other commemorates Harry, himself, who had died four years earlier of cirrhosis of the liver; perhaps he had dived too deeply into his father's capacious North Terrace cellar. On the deconsecration of St Paul's in the early 1980s, the windows went by way of Mr Adams' foundation, Pulteney Grammar School, to the Art Gallery of South Australia, where they are now displayed to advantage.

Adding to the emptiness of Sir Henry's later years were periodic family absences overseas. Sir Henry himself never went abroad again after 1870, though he was fascinated by Canada and donated a book on its geology to Adelaide University Library. Once, after Frank left for a trip to Britain, a pound was found in his bedroom at North Terrace and this was dutifully laid to his account with H.L. & A.E. Ayers by his scrupulous father.

Fred's family settled for the southern-summer months up in their Hills residence, Altnabreac, at Aldgate. He felt unwell in the New Year of 1897, but it passed off quite quickly, and he soldiered on suffering no more than his normal gout. Later in the month, this flared up and he was in no shape to go in to the city office. Doctors were called in, but his family was not alarmed. He was not by nature one to complain. However, complications set in and, with his family around him, Fred died on 1 February, a month short of his fiftieth birthday. The newspapers commented on the great regret that would be felt in Adelaide circles, given his 'genial character'. He was a leader in society and sport, and had been a 'host without peer'. Fred was buried at North Road Cemetery, where his sister Maggie already lay. Many people, from all walks of life, were at the service. Even Frank Rymill, the seemingly more easy-going of the two Adelaide Rymill brothers, said his farewells.

Just a month later, many of the same local personalities and dignitaries – to a large extent the sons of those of Sir Henry's contemporaries who had been in the public eye – met again at the graveside of one of those contemporaries, Sir Henry's good friend, Sir Thomas Elder. The 79 year old's health had been declining for some years. Yet, his passing would occasion 'deep and general regret', claimed the *Advertiser*.

Sir Thomas' death notice appeared alongside reports of the elections for ten delegates to the Convention, shortly to meet in Adelaide, at which a Federal Constitution was to be thrashed out for Australia. On a very low turnout, Charles Cameron Kingston headed the poll, with Sir R.C. Baker not far behind. Slotted between was Treasurer Frederick Holder, who would become the Federal Parliament's first Speaker.

The End of an Active Life

Sir Henry was absent from both Fred's and Sir Thomas' funerals. For an elderly widower, shocked by the unexpected death of his son – the academic, the keen sportsman, the genial host, the lawyer he never quite was himself – that he was not at Fred's funeral is perhaps no surprise. Sir Henry does not even seem to have made it to Fred's sickbed. He may have been in too low a condition himself; he was certainly ill by a fortnight later. It was reported that he had borne up well, outwardly, under the stress of Fred's death. No doubt, in the privacy of his North Terrace home, Sir Henry felt the full force of this new loss. By the SAMA Board meetings of April, it is the 'Acting Secretary', Harry, who is mentioned. By his seventy-sixth birthday, on May Day, Sir Henry did not expect to see the Queen's Diamond Jubilee in mid-year, and certainly not another Christmas.

Sir Henry died early on 11 June 1897. The large bell of the Adelaide Town Hall was set tolling and flags lowered to half-mast. His death notice was carried by the *Register* amid reports of the Diamond Jubilee, and the unimaginable next step in communications: a forthcoming experiment by Marconi of sending a telegraph-message across the Atlantic without the aid of wires – 'bordering on the inconceivable'.

His death marked the passing of a significant link with the early days of South Australia, the *Register* believed, as one of the colony's 'most distinguished public men', inseparable from the story of its development. His Secretaryship of the SAMA had linked him to South Australian industry for over 50 years. As a financier, he was one of the colony's 'safest and acutest'. He had always displayed great caution and penetration, combined with 'reasonable enterprise' and a 'thorough honesty of purpose'. The link with the Savings Bank, the 'Popular Money Box', stretched back over 40 years and, despite his deterioration, he had been re-elected Chairman eight days before he died. His connections with the Gas Company, the Bank of Adelaide, the Botanic Gardens, and the Treasurership of the University, were also rehearsed.

Sir Henry's support for Goyder's revaluations of the pastoral leases, in 1864, was seen as one of the high points of his ministries which, through those years, saw prosperity generally for the colonists. His governments ensured that the colony's debt was kept low for a decade from 1863. The obituarist reckoned his 1875 opposition to succession and stamp duties sensible, in retrospect; in fact, Sir Henry favoured, and later introduced, these duties. The idea that he was over-cautious in government – a mere 'red-tapist' – gave way, in his obituary, to the image of a man who had been remarkably vigorous throughout his political career.

The obituarist referred to the 'dramatic scenes' of 1877, when the leadership of the government was taken from him and handed to William Morgan. In the President's Chair, his 'culture and diplomatic training' had given an added dignity to the office which would be hard to follow. As had been said of a former Speaker of the House of Commons, 'His face and figure fill the eye and his voice charmed and impressed the ear.'

Of Sir Henry himself, the writer remarked:

> His uprightness in all things, his sturdy independence, his varied gifts, his tastes, his dignified courtly bearing and his personal association all combined to invest him with a distinction excelled by few if any of our colonists past or present whose names are written large in the pages of colonial history.

Allowing for hyperbole, and a reluctance to speak ill of the dead, one can still sense that the little lad from the backstreets of a dockyard town had indeed grown into a remarkable personality.

Frank Ayers graciously declined the offer of a State funeral, sure this would have been his father's wish. On Saturday, the twelfth, the 62-carriage cortège started from the North Terrace house, and swiftly swelled to over a hundred. The hearse led the turn, opposite the old and new Parliament Houses, into King William Street, past Imperial Chambers and the Waterhouse Building, the Savings and Adelaide Banks, the Town Hall, and the Treasury Building, which had been the centres of Sir Henry's active life over the previous half century and more. Off Victoria Square, along Grote Street, the procession passed along the route where Sir Henry had trudged, in stifling mourning clothes, behind the gun-carriage carrying Sir Dominick Daly's coffin, a generation before. On this winter's afternoon, mourning black would be welcome.

Sir Henry's might as well have been a State funeral, too, considering the numbers who lined the streets. The West Terrace cemetery was also crowded with people awaiting the coffin's arrival. There Sir Henry was laid to rest, with Anglican rites led by Canon Sunter, in a vault that already held the remains of Lady Ayers, and the small caskets of Charles Coke and Sydney Breaks – the first resting there for nearly half a century already. Another occupant was Lady Ayers' stepsister, Elizabeth Churcher. Frank and Harry led the mourners, with Arthur Lungley and John Bagot – the two sons-in-law – also present. Grandsons were there, in the shape of Sidney Ayers and Walter Bagot. Sir Henry's long-time clerk, James James, also had pride of place. The name of Sir Henry's cousin, William Hayward – Chief Officer of the RMS *Victoria* – was also recorded. A wreath travelling to the cemetery poignantly highlighted the fact that only one of Sir Henry's daughters could mourn his passing: it was from the second Mrs Lungley. Three unnamed female relatives paid their respects in the vault. Speaker, President and Chief Justice Way attended, while Robert Barr Smith and Lance Stirling represented the university. The Civil Service heads were also there. The Gas Company was represented, as was the SAMA, by Frank Rymill. Was Henry Rymill present? Intriguingly, one newspaper records his presence, while the other does not.

The 'private' nature of the funeral precluded the Governor's presence, but just about every other prominent person attended, including all government ministers except Premier C.C. Kingston, who was in London for the Queen's Jubilee. Henry Rymill, his sharp tongue little mellowed with age, seemed to envy Kingston his good fortune in having just chaired the Federal Convention, as the local Premier, and also in handing the colony's congratulations to the Queen, 'a grand slice of good luck'. Mrs Kingston would bring no credit on the colony, as she was 'such an ordinary looking person'.

Someone missing from both Fred's and Sir Henry's funerals was Ernest Ayers. He and his family were abroad, and he must have been wondering whether it was safe for him ever to go away, as he and Barbara were visiting Britain in 1881 when his mother died.

Bequests and Eulogies

A month after Sir Henry's death, the SAMA Board met minus its Secretary of the previous 52 years. Chairman Frank Rymill proposed that the Secretaryship should be taken on by William West – soon succeeded by Mary Elizabeth's eager young husband, Howard Lloyd. The board went on to place on record the

> deep sense of loss sustained by the South Australian Mining Association through the death of their Secretary the late Sir Henry Ayers K.C.M.G. who has since the foundation of the Association so successfully managed the affairs of the Company, and whose untiring energy and zeal in its interests, has placed the Company under a debt of gratitude to him – and the Board deplores the said event which has deprived the Association of his valuable services.

Sam Tomkinson spoke of his long-time friend, and they all recorded their thanks to 'James James clerk to the late Secretary for the efficient manner in which he [had] kept the books, etc., of the Association for the past thirty years'.

Sir Henry's 849 shares were applied for by executors Frank and Harry. There was a very handsome estate to be distributed. The thrifty father, at 30, had counted his pennies up to a proud total of over £5,000. Forty-six years later, Sir Henry had amassed a good 45 times that amount – not to mention the sizeable sum he had entrusted to Maggie and her children, nor any money he may have lost through the collapse of the Commercial Bank.

Sir Henry had made benefactions, including to his brother, John. From the newspaper report, it did not appear as though he had been particularly generous to Josey, leaving her the carriages and horses and just £2,000 for maintaining them. This overlooks a codicil to his will, in which Sir Henry made very generous provision for her, as well as increased legacies to housekeeper Bridget Galvin and James James. Further bequests went to organizations such as the Lady Ayers Homes, the Home for Incurables, the Adelaide Children's Hospital, the Sisters of St Joseph, Schools for the Blind, Deaf, and Dumb, the YMCA and the Church of England Institute. After covering expenses and debts, the residue was to be divided equally between his sons. With Fred already gone, his portion was to be held in trust for his children; Sir Henry was kindly enough to have stipulated that these young people could spend it how they liked and in no way have to account to his executors.

Bachelor Frank faced a major decision: should he continue to occupy the family home? What would one quiet man do with a ballroom and a large formal dining room? It appears that none of the other families wished to continue links with the house. By 1898, Frank was living at the Adelaide Club, pending completion of a house of his own in McKinnon Parade. In North Adelaide, Frank was nearer other family members.

The Legislative Council noted the passing of its longest-serving former member. The Honourable Dr Campbell described Sir Henry as a 'typical product of a large number of men, which only a new country brings into existence'. Despite necessarily being caught up in working with the 'raw material of forces of nature', forging his own commercial and political careers, their distinguished predecessor had leaned also towards the 'advancement of art and cultivation of higher education in the community'. Courteous amid the hottest disputes, he had always struck the speaker with his clarity of thought and expression. Chief Secretary O'Loghlin said he had sat on the Board of the Savings Bank with Sir Henry, who rarely missed a meeting. This was but one example of Sir Henry's assiduity. Sam Tomkinson underlined his 45-year association with his late friend and drew colleagues' attention to the three portraits on the wall opposite. One was of Sir James Hurtle Fisher, one of George Fife Angas and the other of Sir Henry, all three of whom, he added, he had been able to wish a happy birthday on May Day. Sir E.T. Smith, Treasurer of the Old Colonists' Association, had seen Sir Henry's works of kindness in that sphere. In all these areas, he said, Sir Henry had given valuable service to South Australia for which he had gained not a shilling from anyone.

To round off, Sir R.C. Baker, Sir Henry's successor in the President's chair, also claimed personal friendship, and spoke of his predecessor's unfailing courtesy to political supporters and opponents alike. Considering that the younger Baker had taken up – as an oppositionist – where his late father had left off, while Sir Henry had still been active in the chamber, his kind words were particularly gratifying.

Then, it was business as usual. The right Address for the Queen's Diamond Jubilee needed consideration, and the draft Constitution for a federated Australia called for careful scrutiny – paraphrasing Sir William Milne, their Mr Ayers would have had decided opinions on its clauses.

Henry Rymill wrote to Harry Graham in the August after Sir Henry's death, but mentioned not a word about it. He observed that 'the Colonials' had apparently been well treated in the 'dear old Country' and he hoped that their visit would have a beneficial effect upon them. He sounds as though he remained to be convinced, considering Charles Cameron Kingston 'a queer man'.

One wonders what Sir Henry's most recent position was on federation. One cannot imagine that, even as a sick man, he took no interest in the outcome of the Convention which had deliberated a stone's throw from his home. Catherine Helen Spence, the elderly campaigner for electoral reform – particularly for Proportional Representation – had, during Sir Henry's last weeks, become the first woman to stand for election. She fell short of the votes needed to be elected one of the South Australian delegates, but she was by no means bottom of the poll. The man who had spoken forcefully of women's rights, in putting up the Married Women's Property Act a generation before, even if sick, would surely have been interested to see how Miss Spence had fared. It would take a little over three years for there to be an end to colonies and the beginning of Australian nationhood. Sir Henry had said it would not happen in his lifetime.

Ayers in Retrospect

Do today's holidaymakers, dawdling along the seafront at Glenelg in 40°C January heat, make anything of the small plaque labelling some converted holiday apartments as Sir Henry's one-time summer residence, Seafield Tower?

If the Ayers sons had followed their father onto the political stage, as with the Kingstons, the name might have stayed alive, but they led comparatively quiet lives. Even Ayers Rock is receding, while Uluṟu comes forward. Not only has the name reverted but custodianship of the Rock and its environs has been handed back to the Aṉangu people, in whose culture it naturally continues to assume central importance. So how should we remember the public and private man behind the colonial name?

Firstly, of course, Ayers administered a prodigious mining enterprise which served to buttress a struggling British colony when it needed that support most. Indeed, he was characterized as Australia's first mining industry giant. The unmatched achievement of being seven times Premier, 11 times a minister and 12 years President of the Legislative Council deserves recognition, as does his more ephemeral eloquence. Accusations that he was a red-tapist underrate his achievement of myriad incremental financial and government-administration reforms. He also set in train his colony's expansion into the Northern Territory, forestalling Queensland's ambitions and placing South Australia in the best position to maintain its relevance when the technological leap of the world-linking telegraph was accomplished. That, too, he saw to successful completion when the project looked like foundering.

Of course, such achievements represent white man's history, and overlook the dispossession of the Aboriginal peoples of the northern part of Australia once those lands were 'opened up'. No doubt Ayers believed Northern Territory Aborigines' acquiescence in the establishment of a British foothold in their lands could largely be bought, as in South Australia, at the price of doling out flour and blankets – or in his case, gifts of counterpanes.

While Ayers was sufficiently bold to set up the Northern Territory expansion, its initial struggles could partly be laid at his door, too. His Burra 'hands-on' method may have found its limits when applied to the Northern Territory expedition. Finniss had no say in the choice of personnel forming his party; the choice had been that of Ayers' government.

Ayers' political legacy is also measured by his social reforms in the 1870s. He may not have quite enacted Public Education, but he spoke fervently in its support when numbers of prominent contemporaries were unequivocally against the principle. Also, he spoke warmly in support of the right of women to independently own their property after marriage, and did his best to enact that as well – a step on the road to the vote, 20 years later. He showed a compassionate streak in his reforms, adopting the Boarding-Out idea for children of the destitute, moving to end imprisonment for debtors, and seeking to rehabilitate drunkards. Avoiding stigmatizing the first and last of these was also important to him. Perhaps surprisingly, it was the financier and arch-capitalist who spoke up eloquently for the Eight-Hour Day, and legislated the legal framework for trade unions.

The critic can carp that it is all very well but he got neither a Married Woman's Property Act nor Public Education through the parliament. This is true, but if he had not been forced out of office on what might easily rank as peripheral matters, in 1873, no doubt these advances would have been achieved. Given the lack of a disciplined party system, each of his governments fell prey to the manoeuvrings of his political opponents who, very often, found themselves faced with the selfsame inability to carry a programme through. This would not change until the formation of a more rigid party system, with the advent of Labor around 1890, and R.C. Baker's response with his National Defence League which liked to pretend it only consisted of individuals while not constituting a party. Only then was C.C. Kingston enabled to keep his footing solid for most of the 1890s, overseeing votes for women and chairing the Constitutional Convention. This earned him a statue in full Court-Dress, in Victoria Square. By contrast, Ayers only has one of the many pavement plaques on North Terrace.

Given the then-existing limitations on Premieral power, could Ayers be seen as in any way approaching the level of statesman? W.H. Cooper, addressing him from the pages of *The Quiz and Lantern* in 1887, reckoned not. This was a much straighter assessment of Sir Henry's political worth, perhaps, than the warmer pieces written at the time of the subject's retirement from parliament and indeed at his death.

Cooper had been impressed with the politician's dignity in 1877 at the height of the Morgan/Baker attack, especially as R.C. Baker's position had been essentially self-seeking. The two main newspapers had also been intent, maintained Cooper, on getting Colton out of office and so Sir Henry's active political career fell victim to that end. Incidentally, in a somewhat similar vein, ex-Governor Fergusson had written to Sir Henry, believing him a victim of circumstances in his previous losses of government. Cooper believed that, had Sir Henry not been finally pushed out of government – and it was unconstitutional, of course – then perhaps he might have mounted the statesman's pedestal.

Though admiring Sir Henry in so many ways, Cooper could not forgive his acceptance of the knighthood when he had passionately opposed the whole concept of the telegraph. On a clipping of Cooper's article, Sir Henry drew black lines either side of this paragraph, asterisked it, and wrote his own refutation below. It was 'Absolutely untrue'. He said it would have been impossible for him to have spoken against it – in parliament certainly – because the decision had been taken to go ahead with the mammoth project while he was sailing for Home in 1870. In fact, three years before, one member had thanked Ayers for his warm support in parliament for such a project. At that time, opinion in the colony was divided; Henry Rymill for one was very lukewarm, believing the colony should pay off its debt first. A brief article, refuting Cooper's 1887 contention, was published in *The Farmers' Weekly Messenger*, reinforcing Sir Henry's private defence.

Cooper dubbed Sir Henry a conservative but, in saying that, he said there were Conservatives and Conservatives. 'If all our capitalists had the same liberal instincts as you possess, I should not fear the results of the impending conflict between capital and labor. There would be no bloodshed then.' Cooper obviously viewed the rise of the working class, and the probable intransigence of those then able to wield power in confronting it, with the same forebodings raised with Sir Henry by the Reverend Petavel in 1870. Sir Henry's conservatism Cooper reckoned no more than an accident of the times. Had he been born 20 or 30 years later, he would undoubtedly have been a Radical. Indeed, he would have to have been a strange kind of Conservative anyway if Cooper's next claim were true: 'I recognize in you a man of great ability, whose instincts incline towards the traditions of the past, but whose reasoning powers lead him to be almost persuaded to believe in the new doctrine of State Socialism.' Cooper felt the 'pain' of a conundrum: even while believing the foregoing of him, Sir Henry represented 'at least in part' land grabbers and monopolists.

Land grabbers and monopolists; not the most attractive of bedfellows, which leads one to consider the couple of columns in a newsletter of the Historical Society of South Australia, of September 1996, entitled *Sir Henry Ayers: liar and a bastion of the propertied classes?* Geoffrey Manning put together a short collection of newspaper quotes which might serve as a counterweight to all the conventionally nice things said of Sir Henry at the time of his death. He does not give any context, nor discuss them. They are just offered for consideration, examples perhaps of what Robyn Taylor (1997) has described as the darker side of Sir Henry's story. To what do they refer? The first is the line of John Stephens, in the *Register* at the time of the miners' strike, referring to Ayers as grinding the faces of the poor. Now we know that Stephens was virulently anti the nonentities on the SAMA Board, and though Ayers was the public face of the Association – very much in Stephens' face because the Secretary's name was appended to every little advertisement or notice which came into the *Register* office almost daily – his actions were then only at the behest of the directors. Even confronting the miners, a number of the directors were present; Ayers was not alone, and the suggestion for the ensuing lock-out was sent by MacDonald, obviously coming from some of the other, town-bound directors. However, let us not whitewash him too much; that he was a hard man in his younger days there is no doubt. Manning also included a couple of quotes from editorials of John Howard Clark at the height of the New Parliament House crisis of 1877, including the jibe about a small piece of paper to list his achievements. That the dying Unitarian Clark was disappointed by a long-time associate in the little intellectual community, by a man from whom he had expected more, comes through clearly. That James Penn Boucaut was also disappointed in the same man, and was heavily critical of Ayers – not because he had hoped for more from him but, to the contrary, he was keeping him (Boucaut) out of the premier office – should not surprise. Therefore, the contents of a letter to the *Register*, penned by Boucaut in March of the same year, are no surprise either. It shows how short-lived gratitude could be, considering how Boucaut had implored Ayers for support during the investigation into the former's alleged corruption – support that had been duly forthcoming.

One newspaper quote in Manning's article, again from Clark, is probably on the mark: with Ayers in opposition, late in 1875, Clark's line was that Ayers had never done anything to develop South Australia's resources, that he always waited on other people's enterprises in business and politics. Clark attributed this to the fact that Ayers was less a man of original thought than a man of detail. As we have seen, this possibly played some part in Finniss' lack of success in the Northern Territory. It was very difficult for Ayers to escape his upbringing and conditioning. Desk-bound from an early age, with his head in ledgers and letter-books – on Sundays, too – it was perhaps difficult to see the far horizon, let alone beyond it. As to actually 'doing things' while in office, one of the major criticisms of him in the 1870s, in his tussles with Boucaut and Morgan, was that, where they were willing to borrow and spend lavishly – to the tune of £3 million – Ayers was supposed to be a non-borrower. More or less praised for this in his political and actual obituaries, it was not the complete story. His governments did borrow, but moderately. Moderation, caution, were the marks of the man. Of all people, Ayers was not the man to put South Australia's finances at risk.

It seems odd to see in someone once prepared to risk everything on a dash to an experimental new colony the primary element, caution. Risk appears over and done with once the migration step was taken. In this respect, he was unlike J.B. Graham, who risked all on the Burra venture. He, however, then had no family responsibilities and his father's bankruptcy as a precedent, had his luck not held. Henry's luck was that of a beginner, and it held because he had the acumen – the 'nouse', as Mr Adams would have written – to capitalize on it.

Was he as 'tight' as sometimes suggested? He was certainly not a benefactor like Sir Thomas Elder or W.W. Hughes, who could give by the tens of thousands, but then nor was he a money-maker in their league. Giving by the hundreds still placed him in the second rank – of only two – in donations to the nascent university, and produced the Lady Ayers Homes, to the maintenance of which he continued to contribute. Twenty pounds and Sir Henry considered his debt to Portsmouth's Beneficial Society school paid, in supporting, specifically, the girls' section, while £50 started off the fund to provide a fitting monument to Colonel Light. The 40-something Mr Moneybags who would not spare a farthing more than necessary to help a friend, again according to – an admittedly disgruntled – Henry Rymill, gives way to a more open-handed underwriter of Old Colonists fallen on unhappy times. As Honorary Secretary of that association since 1883, he would hear sometimes lengthy recitals of trials and tribulations before giving them something to tide them over. Perhaps the loss of Lady Ayers stirred a greater sense of patience with other people's difficulties than Sir Henry had demonstrated when confronted with sob-stories in London. When the Association's funds on occasion sank low, he topped them up from his own pocket, and even put in a word for those who perhaps had called on its support a little too often.

The relative largesse of his later years was dispensed from a plateau of comfort to which he had ascended over many years. In his younger days he had been a hard task-master of the Burra workforce, always looking for the pound of flesh from worker, officer and agent alike. Yet, by his fifties, if Henry Rymill is anyone to go by, he was nowhere near tough enough. While his monetary donations may not have been extravagant, of his time, Sir Henry was certainly prepared to give most lavishly.

His Greek mythical representation of the SAMA and PCC players, and the amused eye cast about the London Season reveal a man who could put his finger on a character and read a situation as well as he could read himself. The fact that he did not take himself too seriously, loved to indulge in gossip, took an interest in having a well-stocked wine cellar – had an eye for the ladies – and delighted in the company of his daughters and grandchildren, marks him out as no stuffy Victorian-era fuddy-duddy. Evenings playing cards, building with his sons' bricks and dispelling the boredom of a Derby Day traffic-jam with pea-shooters also serve to dispel the latter image. Enjoying good food, music and theatre, giving a song, acting in amateur dramatics or taking a lengthy constitutional, all round out the man who otherwise spent so many hours chained to a desk. Perhaps it is not surprising that he was such a chum of William Paxton, the Burra miners' most popular SAMA Director – the chemist who owned the pubs.

What of the inner man? Ayers' Anglican adherence certainly seemed nominal in his younger days. Yet to step outside the bounds of the established church – even if not established in South Australia – into the radical, reformist nest of Unitarianism was a major movement. The denomination had only been granted toleration as late as 1813 in Britain as its adherents were not only 'heretical' regarding traditional beliefs involving the Trinity, but they advocated for political and social reform. Asa Briggs (1959) notes that Unitarianism was closely linked with Bentham's Utilitarianism and the reformers of the 1820s. The Adelaide Clarks, prominent among the founders of the Wakefield Street congregation, hailed from the Birmingham hotbed of Unitarianism. If Henry and brother-in-law A.J. Baker had crossed over from the Anglicans and William Morgan had come from the Congregationalists, Catherine Helen Spence, and her bank-manager brother, had moved from Scottish Presbyterianism – Miss Spence giving the Reverend Crawford Woods a trial before converting. Even Woods himself had formerly been a Presbyterian in Ireland. Henry was therefore far from alone in choosing Unitarianism as a more comfortable spiritual home. Did he convert through his friend among Burra directors, 'Mr Billy' Sanders – who bankrolled the church building? Or, did the traumatic loss of one-year-old Sydney, a year or so later, steer Ayers into looking for something more satisfying? He found it among the humanist – almost secular – Unitarians. Their appreciation of high culture, the general intellectual standard of their small community, all seem to fit with what we can glean of Ayers himself. Undoubtedly, whatever reformist transformation Ayers was undergoing himself, it could only have been magnified by this circle, coming to fruition through his ministries of the 1870s. Furthermore, if the Unitarians could see a God as the originator of evolution, no wonder Ayers could rush down for lunch with Thomas Huxley. Science was, for him, almost a god when it came to agriculture, or even applied to the poor old Burra. He wrote from London of wind-speed gauges and soil thermometers, and was amazed by both the brute force and precision capabilities of steam power that he witnessed there. The university, on whose Council he served as Treasurer, was the first in Australia to bestow science degrees, not to mention degrees on women.

With education, we end where Henry's story began. His father, substantial, unspectacular William Ayers, who had risen not at all beyond a safe and comfortable position in the Royal Dockyard, had gifted his son a few precious years' schooling. This opportunity Henry had grasped with both hands, at the age of 19, in a way that meant he never saw his parents again. Only once, briefly, does he seem to have contemplated settling back Home. As he dug himself deeper and deeper into Australia's commerce and politics, he recognized he had cut his suit of a colonial cloth, and would wear it – proudly.

APPENDIX A:
WEIGHTS, MEASURES AND MONEY

Any attempt to estimate modern equivalents to the prices and salaries of the mid to late nineteenth century raises a wide range of possibilities. For example, using the website EH.Net (Economic History Services) at Miami University, a comparison based on changes in average earnings between then and now gives a certain figure, while a comparison based on average prices gives a figure ten times smaller. Using the average earnings calculation, £1 (a pound) in 1845 would have been worth about £770 in 2008 – the most recent comparison available. Using average prices, Miami suggests more like £71 in 2008 terms. Rather than my stating one or other comparison in the text, I stick to actual salary-figures and prices of the day and leave it to the reader to use whichever version he or she prefers.

To add a little more complexity, the Miami website is only making comparisons between earnings or prices within Britain, then and now. Although sterling currency was also used in the colony of South Australia, there could be a small difference between the value of the pound at Home and in the colony due to bank commissions on sending or receiving bank bills – a rudimentary exchange rate, one might say.

Also, as time goes on, the difference between values reduces, e.g. the £1 by 1870 would be worth more like either £590 or £71 in 2008 terms. It had dropped in value as a result of the 1848 Revolutions, for example, and after the financial crash of the London market in 1866. By Sir Henry's death, in 1897, each pound he left would only equal about £470 of ours, or £85 (depending on whether you reckon by average earnings, or retail prices). In his will, then, the actual £225,000 Sir Henry did leave would be the equivalent of either something over £105 million or nearly £20 million. He was not short of a bob (a shilling) or two, either way!

Taking one example using the average earnings comparator, Henry Ayers' seeking for a rise, in 1849, to a salary of £600 per annum might equate to a salary in 2008 terms of something in the mid-£450,000 range. In today's Australian dollar terms, at say $1 = 60 British pence, that salary might be $750,000. As the then Secretary of a thriving mining company, a salary of that order sounds more likely than one a tenth of the size, had the average prices comparator been used.

As the old money disappeared from use in Britain in the early 1970s, and from Australia in the mid-1960s, I might explain the old system and the way of writing it. The pound was divided into 20 shillings – denoted by an 's' – and each shilling was divided into 12 pence, denoted by a 'd' – both the '£' for pound, and the 'd' for pence are derived from Latin originals. So, the pound was divided up into 20 × 12 pence, i.e. 240 pence to the pound. The simplified, decimal 100 new pence – 100p to the pound – meant the new penny was worth 2.4 old pence. One can see that the modern 50p is equivalent to the old 10 shillings, which would have generally been written 10/-, or 10 shillings and no pence. To equate to £1.50p, one would have written £1/10/-. 'Fancy' things were often priced in guineas, which was a posh way of charging a little bit extra for them – it might be an item like a smart hat, or a term of ten piano lessons. The guinea was 21s, or £1/1/-, or simply 1gn. A sovereign was a gold coin of the value of £1 – very popularly made up into pendants of recent years – while a crown was worth a quarter of £1, i.e. 5/-.

I have kept to distances in miles, e.g. the Burra mine lay about 100 miles north of Adelaide. The easy equivalent in kilometres is 50 miles = 80 kilometres: divide the mileage by five and multiply by eight to get the number of kilometres. Smaller measures I have kept in yards, feet and inches. The yard is a little shorter than a metre. It divides into three feet, and each foot – like shillings breaking down into pence – is made up of 12 inches.

As with depth measurements at sea, the depth of a mineshaft was reckoned in fathoms – 6 feet in a fathom, or 2 yards. When they talked of going down a mine shaft to the '20', for example, they were talking of a depth therefore of 120 feet, or 40 yards – the best part of 40 metres.

Weights are a little more complicated: the ton is not much different in weight from the metric tonne but, like the £, it divides into 20 parts, called hundredweights, and written cwt – c deriving from the Latin for 100. We could

APPENDIX A: WEIGHTS, MEASURES AND MONEY

go on, dividing the hundredweight into quarters, themselves divided into two stones, and each stone divided into 14 pounds – written lb. *They* could go further into 16 ounces – or oz – still the unit in which gold is priced internationally. Tons carry us a long way in Ayers' story, except the only snag is, in the copper industry with which he was closely linked, the ton was not reckoned at 20 cwt but, rather – like the guinea in money – it was reckoned as 21 cwt. This was not because copper ore was in any sense posh, but it simply made some allowance to smelters for their costs between purchase of ore and the sale of pure copper.

APPENDIX B:
THE WESTMINSTER SYSTEM

For those unfamiliar with the Westminster parliamentary system of government, as found in Britain and Australia, it may be as well to understand how a policy is turned into law, as an Act of Parliament. The chosen policy is shaped into a bill, drawn up by someone with a legal mind, and is introduced by a Member of Parliament, who seeks a vote of support that the bill should be read. Bills can be put forward as government measures, or come from the backbench, as a Private Member's Bill, and might even result from a backbencher's taking up the cause of petitioners from the general public to the parliament. A government may or may not allow time for such bills.

Three readings of the bill take place. The real work of discussing it fully, hacking it about – through amendments to its clauses and its phraseology – is usually done at the Second Reading stage and when the whole House sits as a Committee. When everyone has had his or her – though not in Ayers' day – say, the version that comes out is reported and ordered to be printed in readiness for the Third Reading. If all remains well, the House passes the bill and sends it for the other house, in a bi-cameral parliament, to do the same thing, and see if it can come up with any further improvements – or, if a government does not command a majority in an upper house, a bill may be emasculated by a determined Opposition or even flatly rejected. If it has simply been agreed to or in some way improved, it will be sent back to the first house to see if its members agree that improvements are indeed improvements and, when everyone is happy, the bill has passed both houses. For it then to become law, it requires the stamp of approval from, in the British context, the monarch – at present, Queen Elizabeth II. Once she has given her assent, the Act becomes law.

In Henry Ayers' time, when South Australia was a British colony – or Province – most Acts could be assented to by the monarch's representative, the Governor. If the Governor felt that a certain Act might be controversial, or might even go against the principles of a similar Act of the Parliament in Westminster, he could reserve a decision and send the Act to London to find out if he should assent or not. This reserving of Acts for London's consideration no longer obtains for state Governors. However, the Acts of the Federal Parliament in Canberra are still assented to by the Queen's representative, the Governor-General or, if the Queen happens to be visiting, she can do a spot of assenting in person.

APPENDIX C:
THE FATE OF THE SAMA

What happened to the South Australian Mining Association? Its fiftieth anniversary report was prepared by Sir Henry in 1895. By the 1896 AGM, so many friends' faces had disappeared: Sir John Morphett, both Blyths, Paxton, of course, and Sir William Milne. John Beck, still a survivor, was attempting to sell the entire SAMA property for £55,000 in London, without success. After Sir Henry's death, it was another four years before brighter prospects of a sale appeared: T.B. Gall, erstwhile partner of Frank and Fred Ayers, was negotiating a sale for the Burra Burra Syndicate, for £10,000 down and £12,000 worth of fully paid up shares in the new company. Sir Henry's son-in-law, John Bagot, seconded by Ernest, proposed that Chairman Frank Rymill treat with Gall and the syndicate, preferring £6,000 down and £24,000 in new shares, with the stipulation that the mine must be worked within three years.

The firm had become even more of a family one, the year before, when Harry's son-in-law, Howard Lloyd – married to Mary Elizabeth, the babe so much in the Ayers' thoughts during the London visit – became Secretary. Here was a young man very much in the Sir Henry mould as far as business acumen goes, more so perhaps than any of Sir Henry's sons themselves. Given time, Howard Lloyd would earn a knighthood, too. Roughly quarterly SAMA Board meetings continued, though amounting to not much more than paper shuffling, in which more descendants and relatives took part; grandson Walter Hervey Bagot and Hurtle Morphett took their turns. Frank Rymill sailed on as Chairman and it was he, Morphett and another Bullock family member who were advised – by Frank Ayers' erstwhile law pupil, W.J. Isbister – on the winding up of the company, around the outbreak of the First World War. Morphett's Engine had ground to a halt the best part of 40 years before and now, at last, the hot air in the board room finally went cold as well. The Public Library, Museum and Art Gallery made a polite appeal for the papers telling of the early days, and the last act took place where it had begun all but 70 years before: the office furniture was sold at Industrial Buildings – The Auction Mart – by executor Ernest Ayers, his brother Harry, named as co-executor, of course long dead. A sum of £12/5/- was realised, and it would be nice to think that the office table in Smart and Bayne's premises, bought at £1/15/- for Henry Ayers to work at all those years before, was among the lots.

BIBLIOGRAPHY

Primary Sources

Ayers, F.R. (1853). Letter to his teacher Richardson Reed, 11 June, Adelaide: St Peter's School archives.

Ayers, H. Papers, photographs, *Memorabilia*, and scrapbooks. Includes letters from: Sir Henry Parkes, J.P. Boucaut, C. Todd, Sir James Fergusson and other governors, Lady Musgrave, Tom Slade, AMP, Bank of Australasia; condolence letters, etc. State Library of South Australia, PRG 67 box 1-16, PRG 67, 19/2. Also photograph album PRG 67, 19/3, accounts kept on intercolonial conference trips PRG 67, 39 and 41, programmes PRG 67, 42.

———. (1870). Letters to his sons, Adelaide: Photocopies at Ayers House, North Terrace.

Ayers, H.L. and Ayers, A.E. Papers, including some of Sir Henry's, discovered in archives of successor firm, Evans and Ayers. Majority in the State Library of South Australia.

Cawthorne, W. Collection of William Cawthorne's paintings, Sydney: Mitchell Library, PXD 30, 39, 40, 42, PXA 1490.

Coke Papers. Letters from the Janes Coke, Snr. and Jnr., Henry Simmons Coke, Captain W.L. Powell, Henry Ayers, Leach, Richardson & Co, and Viscount Adare to James Charles Coke in Adelaide; bills of lading, miscellaneous jottings and workings out by J.C. Coke, State Library of South Australia. PRG1114/1, 2, 3, 4, 5.

Gill, S.T. Collection of the watercolours of S.T. Gill, Art Gallery of South Australia.

———. Pictures Index, vols 16 and 20, State Library of South Australia, PRG 1336.

Graham Papers. John Benjamin Graham's journal: early experiences in Adelaide and his return to living in Britain and Germany; journal of his South Australian visit in 1858; his stepfather John Adams' journal; letters to Graham from Henry Ayers, 1848–70; letters of Henry and Frank Rymill to J.B., Louisa, Louisa Maude and Harry Graham; records of the Canowie sheep station; letter of 'Eureka'. Microfilmed in 1961 by permission of M.L. Graham, 25, Park Close, Ilchester Place, London, W14. The State Library of South Australia, PRG 100 and the Mitchell Library, Sydney FM4 1537/8/9/40.

Hansard record of Debates and Reports to the Parliament of South Australia (State Library of South Australia and State Archives. Online. Available www.foundingdocs.gov.au/item.asp?dID=6

Hanson, Richard Davies, letter to Governor Daly, requesting a knighthood. South Australian State Archives GRG 24, Series 90/109.

Lloyd, E. (1844–6). *A Visit to the Antipodes – with some reminiscences of a sojourn in Australia*, London: Smith, Elder & Co.
Lockett, W.H. (1841). Letter to his sister, Elizabeth Churcher, April (in private hands).
Parliament of New South Wales (authorised by). *Documenting a Democracy*, New South Wales Index to Parliamentary Debates, vol I, the 1st to the 5th Parliaments. Derived from the *Sydney Morning Herald*. National Archive of Australia.
Patent Copper Company and successor English and Australian Copper Company records: journals, ledgers and cash books of the former and a bound copy of the annual reports of the latter, 1852–74, incorporating extracts from the letters of the South Australian managers to the London board. State Library of South Australia. BRG 30. (1) ledgers (4 vols), (2) journals (17 vols), (3) cash books (8 vols), Annual Reports ZPER 622.343a.
Petition for honours for Colonel Hamley and various South Australian figures (1869), State Records GRG 24, 86/13.
Pollock, A.J. (1996) 'The Pollock Letters, 1850–52', *Journal of the Historical Society of South Australia*, 24, pp. 131–42. State Library of South Australia.
South Australian Banking Company. Letters from E.J. Wheeler, London Manager, to Edward Stephens, Adelaide Manager. SA/8/3 from 13 September 1844 to 18 May 1847, SA/8/4, letters Wheeler to Stephens from 20 May 1847 to 3 February 1852, SA/19/6, letters Wheeler to Stephens, 25 March, 31 March and 15 May 1848 and letter, Schneider and Co to Stephens, 13 May 1848. Records now in the archives of the ANZ, Glen Waverley, Victoria.
South Australian Mining Association records: correspondence books, letter-books between the mine managers and the Adelaide city office, minutes of AGMs and Half-Year meetings, minutes of meetings of the board of directors – both draft and fair-copy – deeds of settlement, details of leases of Association property in the township of Kooringa to miners, tradesmen and others, rough notes, etc. State Library of South Australia. BRG 22 Series list. (1) minutes of shareholders' meetings, (2) minutes of directors' meetings (6 vols), (3) minutes of directors' meetings in rough (3 vols), (4) letters of the Secretary to mine officials (5 vols), (5) General Letter Book (3 vols), (6) Superintendent's letter book, 1847–9, (8) Mine Accountant's letter book (4 vols), (41–47) miscellaneous matters, Thomas Edwards' letter, etc., (50) Ledger: record of shareholders in Adelaide and London from April 1847, Special list, (61) letters, including from Henry Rymill to Henry Ayers, (63) notes on the discovery of the Burra Burra mine, Minutes of the Provisional Committee, etc., letters of Ferdinand von Sommer to Henry Ayers, etc., and the 1847 Deed of Settlement.
Street directories, Adelaide, of the 1870s. South Australian State Archives.
University of Adelaide. Minutes of the University Association, 1872–74, and Minutes of the Council of the University of Adelaide from November 1874. University of Adelaide Archives.
Volunteers lists. State Archives of South Australia. GRG 149/1; GRG 24 Series 51.
Walters, G.S. Letters regarding the Bullion Act of 1852. South Australian State Records. GRG 24 Series 90/7.
Younghusband Papers. Cash book of the firm of William Younghusband, Jnr. of Adelaide. State Library of South Australia. PRG 34.

Newspapers of the Day

The Australasian Sketcher, The South Australian Register, The Observer (SA), *The South Australian Advertiser, The South Australian, The South Australian Gazette and Colonial Register* (later *The South Australian Gazette and Mining Journal*), *The South Australian Government Gazette, The Mercury and South Australian Sporting Chronicle, Adelaide Punch, Illustrated Adelaide Post, The Quiz and Lantern, The South Australian Satirist, The Monthly Almanac and Illustrated Commentator, Adelaide Miscellany*, 1848–9, *The Free Press, The Burra Record*, the *Sydney Morning Herald*, the *Cambrian, Illustrated London News, The Times* (of London), *The Bombay Gazette*

Secondary Sources, South Australia

Abbott Index of births, marriages and deaths in the Adelaide newspapers. State Library of South Australia.
Aldersey, A.D. (1974). *Pastoral Pioneers of South Australia*, 2 volumes. Compiled from the archives of the South Australian Library. Originally R. Cockburn's compilation from a series of 100 articles published in *The Stock and Station Journal*, 1923. Blackwood, SA: Lynton Publications Pty Ltd.
Auhl, I. (1969). *Burra Sketchbook*, Adelaide: Rigby Ltd.
——. (1986). *The Story of the 'Monster Mine': the Burra Burra Mine and its townships 1845–1877*, Hawthornedean: Investigator Press.
Auhl, I. and Marfleet, D. (1975/1988). *Australia's Earliest Mining Era*, Adelaide: Axiom.
Austin, J.B. (1863). *The Mines of South Australia*, Adelaide: Platts, Wiggs, Dehane, Howell, Rigby, Mullett.
Australian Institute of Mining and Metallurgy (1994). *Sir Henry Ayers*, Centenary commemorative pamphlet of the Australian Institute of Mining and Metallurgy. Copy in the State Library of South Australia.
Ayers, L.L. (1946). *Sir Henry Ayers and his Family: a pamphlet for the Pioneer Families Association*, copy in the State Library of South Australia.
Bagot, Walter H. (1957). *Foundation Members of the Adelaide Club, 1863–1864*, Adelaide.
Bannear, D., assisted by Annear, R. (1990). *The Burra Smelting Works: a survey of its history and archaeology*, Burra, SA: Report for the District Council of Burra, SA.
Blacket, Rev. J. (1911). *History of South Australia: a romantic and successful experiment in colonisation*, 2nd edn., Adelaide: Hussey & Gillingham.
Brown, J. and Mullins, B. (1977). *Country Life in South Australia*, Adelaide: Rigby.
——. (1980). *Town Life in Pioneer South Australia*, Adelaide: Rigby.
Bull, J.W. (1884). *Early Experiences of Life in South Australia*, Adelaide: E.S. Wigg & Sons.
Burra Storekeeper's Cottage Survey, commissioned by J. Carpenter. In private hands.
Carter, J.M.T. (1996). *Burra, 1845–1851: a directory of early folk*, Victoria: The Shalimar Press, La Trobe University.
Carter, J.M.T. and Cross, R. (1997). 'Success and Failure: earliest attempts at the commercial smelting of the "Monster Mine's" copper ore in the Province of South Australia', *Journal of the Historical Society of South Australia*, 25, pp. 18–34.
Castles, A.C. and Harris, M.C. (1987). *Lawmakers and Wayward Whigs: government and laws in South Australia, 1836–1986*, Adelaide: Wakefield Press.
Chaput, D. (1984). 'The Burra Burra Question: to ship or to smelt?' *Journal of the Historical Society of South Australia*, 12, pp. 60–75.
Cockburn, R. (1908). *What's in a Name?*
Couper-Smartt, J. and Courtney, C. (2003). *Port Adelaide: tales from a commodious harbour*, Adelaide: Friends of the Port Adelaide Maritime Museum Inc.
Cumming, D.A. (1982). *Notes on the Burra Mine and Smelter*. For the country weekend of the South Australian Division of the Institution of Engineers, Australia. State Library of South Australia, Mortlock collection 338.274 309 942 32. C 971.
Cumming, D.A. and Moxham, G.C. (1986). *They Built South Australia: engineers, technicians, manufacturers, contractors and their work*, published by authors.
Davies, M. (1977). 'Bullocks and rail: the South Australian Mining Association, 1845–70', *Australian Economic History Review*, XVII: 2, pp. 151–65.
——. (1983). 'Copper and credit: Commission agents and the South Australian Mining Association, 1845–77', *Australian Economic History Review*, XXIII: 1, pp. 58–77.
——. (Unpublished). 'Leakage and Boomerangs – dividends, consumption and investment: John Benjamin Graham and his South Australian mining fortune'.
de Meyrick, R. (2003). *Rymill*, Railmac Publications.
Drew, G.J. (1986/1990). *Discovering Historic Burra*, National Trust of South Australia, Burra Branch.
Dutton, F.S. (1846). *South Australia and its Mines*, London: T&W Boone.
Eade, M. (1971). *Catherine Helen Spence*, MA thesis, Australian National University. Copy in the library of Flinders University.

Elder, D. (1984). *William Light's Brief Journal*, Adelaide: Wakefield Press.
——. (1987). *The Art of William Light*, Adelaide: The Corporation of the City of Adelaide with Wakefield Press.
Finniss, B.T. (1886). *Constitutional History of South Australia, 1836–1857*, London.
Glenelg City Council (1979). *Historic Glenelg: a history of the city of Glenelg to June 1979*, incorporating the book compiled by W.H. Jeanes, OBE, in 1935. On the authority of Glenelg City Council, Adelaide.
Glover, C.R.J. (1939) *Church of St. John the Evangelist, Halifax Street, Adelaide: centenary souvenir*, revised by H.E. Fuller, Adelaide: Hunkin, Ellis and King.
Gooden, G.W. and Moore, T.L. (1903). *Fifty Years' History of the Town of Kensington and Norwood: July 1853 to July 1903*, Adelaide: Webb and Sons, Printers.
Gosse, F. (1981). *The Gosses: an Anglo-Australian family*, Canberra: Brian Clouston.
Government of South Australia (1998). *Discovering South Australia's Mining Heritage Trails*, Adelaide: Primary Industries and Resources South Australia (booklet).
Harcus, W.H. (ed.). (1876). *South Australia, its History, Resources and Productions*, Adelaide: Cox, Government Printer.
Hindmarsh, F.S. (1995). *Powder Monkey to Governor: the life of Rear Admiral Sir John Hindmarsh, Kt, RN, 1785–1860*, Carnarvon, WA: Access Press.
Hoad, J.L. (1999). *Hotels and Publicans in South Australia, 1836–March 1993*, 2nd edn, revised.
Hodder. E. (1893). *The History of South Australia, from its foundation to its Jubilee*, 2 vols. London.
Hodge Index to passenger lists published in South Australian newspapers of arrivals and departures from interstate and from New Zealand, 1837–1859, State Library of South Australia.
Howell, P.A. (2002). *South Australia and Federation*, Adelaide: Wakefield Press.
Jaensch, D. (ed.) (1986). *Flinders History of South Australia*, Adelaide: Wakefield Press.
Jones, W. (1999). *Cousin Dai and Cousin Dilys*, paper delivered to the biennial Cornish festival, 'Kernewek Lowender', at Wallaroo, South Australia.
Kingsborough, L.S. (1965). *The Horse Tramways of Adelaide and its suburbs, 1875–1907*, Adelaide: Adelaide Libraries Board of South Australia.
Lock-Weir, T. (2005). *Visions of Adelaide, 1836–1886*, Adelaide: Art Gallery of South Australia.
Manning, G.H. (1990). *From Aaron Creek to Zion Hill: the place names of South Australia*. An update of his 1986 *Romance of Place Names of South Australia*, Modbury, SA: Gould Books.
——. (1996). *Sir Henry Ayers: liar and a bastion of the propertied class?* Historical Society of South Australia Newsletter, September 1996.
Marsden, S., Stark, P. and Sumerling, P. (eds.) (1990). *Heritage of the City of Adelaide: an illustrated guide*, Adelaide: Corporation of the City of Adelaide.
Morrison, W.F. (1890). *The Aldine History of South Australia*, Sydney and Adelaide: Aldine Publishing Company.
National Trust of South Australia (1984). *Ayers House Story*, Alfred Dunhill (Aus) Pty Ltd.
Newman, J. (1981). 'Emigration from nineteenth century Wales with particular reference to South Australia', BA Honours thesis (unpublished). Flinders University.
Payton, P. (1994) *The Cornish in South Australia: their influence and experience, from mining to assimilation, 1836–1938*, Redruth, Cornwall: Institute of Cornish Studies.
Perry, D.M. (1992). *Sir John Morphett: a South Australian of distinction*, South Australia: The Cummins Society, Inc., assisted by the West Torrens Council.
Pike, D. (1957). *Paradise of Dissent*, London: Longmans, Green & Co.
Potts, L. (ed.), for the Potts Family History Committee (2004). *Frank Potts of Langhorne Creek, his children and grandchildren*, Langhorne Creek: Potts Family History Committee.
Prest, W. (ed.). (2001). *Wakefield Companion to South Australia*, Adelaide: Wakefield Press.
Radford, R. *Louis Comfort Tiffany: the Adelaide windows (incorporating The Tiffany Studios by R. Reason)*, Adelaide: Art Gallery of South Australia (brochure).
Richards, E. (1978). 'William Younghusband and the Overseas Trade of Early Colonial South Australia', *South Australiana – a journal for the publication and study of South Australian historical and literary sources*, 17/2, September 1978 edition, State Library of South Australia, 994 2a.
Sexton, R. (1984). *HMS Buffalo*, Magill, South Australia: Australian Maritime Historical Association.

258 BIBLIOGRAPHY

Stow, J.P. (1883). *South Australia: its history, productions and natural resources. Prepared for the Calcutta International Exhibition*, Adelaide: E. Spiller, Government Printer.
Stratton, J. (ed.) (1986). *Biographical Index of South Australians, 1836–1985*, 4 volumes, Adelaide: South Australian Genealogy and Heraldry Society, Inc.
Taylor, R. (1997). *A History of Ayers House, its Users and Uses*, Adelaide: National History Trust of South Australia.
Thomas, N.H. (1986). *Andrew and James Thomas family: 'Our Ain Folk'*, Mount Gambier, SA: Thomas Family Reunion Committee.
Whitelock, D. (1977). *Adelaide, a History of Difference, 1836–1976*, St Lucia, Queensland: University of Queensland Press.

Australian History: General

Anon. *Ballarat Star*, 6 August 1883. Memoir of *William Meirion Evans*, Manuscript, Welsh Church, Melbourne and National Library of Wales, Aberystwyth. Facs 680.
Blainey, G. (1993). *The Rush that Never Ended: a history of Australian mining*, 4th edn., Melbourne: Melbourne University Press.
Castles, A.C. (1982). *Australian Legal History*, Sydney: The Law Book Company Limited.
Clark, M. (1987). *A Short History of Australia*, 3rd edn, New York: NAL Penguin Inc.
Davidson, G., Hurst, J. and Macintyre, S. (2001). *Oxford Companion to Australian History*, revised edn., Oxford: Oxford University Press.
Flannery, T. (2000). *Terra Australis*, Melbourne: Text Publishing.
Lloyd, L. (1988). *Australians from Wales*, Caernarfon, Gwynedd Archives and Museums Service.
Phillips, V., Latta, D. and Smith, R.G. *New Ways in an Ancient Land*, Sydney: Bay Books.
Serle, P. (1949). *Dictionary of Australian Biography*, Sydney: Angus and Robertson.
Stannage, C.T. (ed.) (1981). *A New History of Western Australia*, Perth: University of Western Australia Press.
Toft, K. (2002). *The Navigators*, Sydney: Duffy & Snellgrove.
Ward, R. (1987). *Finding Australia*, Richmond, Victoria: Heinemann Educational Australia.

South Wales: The Copper Trade, etc.

Banks, A.G. (1984). *H.W. Schneider of Barrow and Bowness*, Kendal: Titus Wilson.
Davies, M. (1998). 'Shipping freight costs: South Australian copper and copper ore cargoes, 1845–1870', *The Great Circle, Journal of the Australian Association of Maritime History*, 20/2, pp. 90–119.
Gabb, G. Lower Swansea Valley Factsheets 6 (Later copper works) & 7 (The rise and fall of the copper industry), Swansea Museum. Archives, Swansea Library.
Jones, W.H. (1922). *History of the Port of Swansea*, Carmarthen.
Rees, R. (2000). *King Copper*, Cardiff: University of Wales Press.
Toomey, R. (1979). 'The Vivians, Swansea copper smelters', unpublished PhD thesis, University of Wales (copy in library of Flinders University).

General Historical Background

Anderson, F.E. *A Visit to St Mary's Church Alverstoke, Gosport, Hampshire*, (brochure.)
Ballard, A. (possibly author, unclear). (1974). *The Parish Church of St Mary, Portsea: a brief history and guide*, Portsmouth: Printed by Coomer.
Bloomfield, P. (1961). *Edward Gibbon Wakefield*, London: Longmans, Green & Co.
Briggs, A. (1959). *The Age of Improvement*, London: Longman.
Darwin, F. (1995, orig. pub. 1902). *The Life of Charles Darwin*, London: Senate, an imprint of Studio Editions Ltd.

Desmond, A. (1997). *Huxley*, London: Penguin Books.
Feuchtwanger, E. (2000). *Disraeli*, London: Arnold.
Foot, P. (2005). *The Vote: how it was won and how it was undermined*, London: Penguin.
Gibbs, V. (ed). (1910). *Cockayne's The Complete Peerage*, London: Doubleday and Howard de Walden, Geoffrey H. White, G.H. White and R.S. Lea/London: The St Catherine Press Ltd.
Hinde, W. (1973). *George Canning*, London: William Collins & Sons.
Hobhouse, H. (1983). *Prince Albert: his life and work*, London: Hamilton.
Mosley, C. (editor-in-chief). *Burke's Peerage, Baronetage and Knightage, Clan Chieftains and Scottish Feudal Barons, 107th edition*, Wilmington, Delaware: Burke's Peerage and Gentry LLC.
Middlemas, K. (1972). *The Life and Times of Edward VII*, London: Weidenfeld and Nicholson Ltd and Book Club Associates.
O'Brien, P. and Quinault, R. (eds.) (1993). *The Industrial Revolution and British Society*, Cambridge: Cambridge University Press.
Seton-Watson, R.W. (1935). *Disraeli, Gladstone and the Eastern Question*, New York: Macmillan.
Thompson, D. (1996). *Europe Since Napoleon*, London: Pelican Books.
Van der Kiste, J. and Jordaan, B. (1984). *Dearest Affie: Alfred, Duke of Edinburgh, Queen Victoria's Second Son, 1844–1900*, Gloucester: Allan Sutton.
Warner, O. (1965). *Portsmouth and the Royal Navy*, Portsmouth: Gale and Polden Ltd.
Weintraub, S. (1987). *Victoria: portrait of a queen*, London: Unwin Hyman.

Comparative values of the pound sterling at various points in the nineteenth century, reckoned from average earnings and from the RPI, with that of 2008. EH.net (Economic History Series). Miami University, Wake Forest University. Available online at http://eh.net/hmit. Accessed 24 April 2010.

INDEX

Adams, John 22–31, 35, 37–41, 43, 46–7, 49, 52–3, 59, 65, 72, 76, 125, 165, 235, 244, 255
Adams, Mary 24–30, 37–9, 49–50, 52–3, 65, 72, 76
Adelaide, Bank of 119, 128, 236
Adelaide, Bishop of 25, 44, 50, 140, 165, 206
Adelaide, City of 5, 10, 14, 16, 19, 23, 28–30, 35, 37, 43, 56, 58, 62–4, 67–9, 71, 83–4, 86, 90–1, 96–7, 109, 112–14, 121, 127–9, 131, 144–5, 152–4, 157–8, 162–3, 166–7, 170, 175, 178, 181, 185–6, 192, 194, 210, 215–17, 221–5, 228, 236
Adelaide, Federal Convention at 1897 236, 238, 240, 242
Adelaide, International Exhibition 1881, 220, and 1887, 226–7
Adelaide, Mayor of 10, 12, 215, 220
Adelaide, Port 14, 22, 32–3, 35–6, 39, 43, 45, 47, 62, 84, 88, 90, 112, 116, 142, 189–90, 208
Adelaide, Queen 5
Adelaide, University of 191, 205, 220, 225, 235–6, 238, 245–6
Admella, steamship 90, 96, 98, 207
Albert, Prince Consort 63, 137
Albert Edward, Prince of Wales 160, 211, 220
Albert Victor, of Wales, Prince 220
Alexandra, Princess of Wales 163, 166
Alfred, Duke of Edinburgh, Prince 136, 141, 144, 151
Alhambra, The, London 163
Allen, William, Captain 22, 59, 64
Anangu people, The 241
Angas, George Fife 5, 83, 240
Apoinga copper works 32, 42, 45, 51, 62

Appleton, brig 35–6
Auhl, Ian 15, 24, 56, 59, 81
Australian Mutual Provident (AMP) 205, 229–30
Australian Natives' Association 6
Australasia, Bank of 16, 34, 75, 108, 118, 120, 135, 184
Austro-Prussian War 128
Ayers, Ada Fisher (née Morphett) 128, 145, 152, 173, 215–16, 223, 235
Ayers, Agnes Marion (later Mann) 209
Ayers, Amy Josephine (later Cowle) 192, 223
Ayers, Anne (later Lady, née Potts) 4, 6–9, 11–12, 17, 30, 38, 48–50, 65, 73, 87–8, 92, 101–2, 106, 114, 122, 124–6, 157, 160–1, 163, 168, 177, 189, 206, 219, death of 221–3, 228
Ayers, Arthur Ernest 73, 87, 157, 161–2, 168, 170, 172, 192, 206, 215, 220–1, 227, 234, 238, 252–3
Ayers, Barbara (née Milne) 215, 234, 238
Ayers, Charles Coke 50, 91, 237
Ayers, Elizabeth (later Ellis, 'Betsey') 9, 47, 119, 159
Ayers, Elizabeth (née Breaks) 2, 9, 101
Ayers, Erlstoun Barbara 234
Ayers, Evelyn Cameron (née Page) 160, 169–70, 176, 194, 209
Ayers, Frank Richman 12, 39, 49, 70, 87, 92, 103, 110, 113–14, 122, 128, 131, 138, 145, 157–8, 160–1, 163, 166, 169, 173, 176, 201, 203, 208, 212, 215–17, 235, 237–9, 252–3
Ayers, Frederic 39, 87, 114, 122, 128, 131, 138, 158–9, 163, 170, 175–6, 194, 203, 208–9, 220, 224, 235, 239, 252
Ayers, Frederic Gordon 176–7

INDEX 261

Ayers, Genevieve 219
Ayers, George 145
Ayers, Harry Cecil 215, 219
Ayers, Harry Lockett 12, 39, 69–70, 87–9, 101, 120, 122–3, 125, 127–8, 130, 138–9, 145, 152–3, 155, 157–8, 160–1, 163, 166, 169, 172–6, 191–3, 201, 207, 215–17, 220, 223, 235–6, 238–9, 253
Ayers, Henry (later Sir) and Aborigines 113, 142, 178, 241, and agriculture 10–11, 136, 143, 147, 163–4, 172, 194–5, 204–5, 217, 246, and alcohol 21, 23, 52, 62, 82, 88, 121, 158, 161–4, 188, 190, 204, 217, 235, 241, 245, and Australian federation 141, 167, 192, 211, 229, 240, and Boarding Out 187–9, 241, and charities 3, 169, 189, 206, 224, 239, 244–5, and civil service reform 118, 194, and education 3, 22, 59, 82, 160–1, 163–4, 171–2, 187–8, 191, 194, 201, 205, 220, 225, 236, 239, 241–2, 245–6, and finance 13, 55–6, 64, 66–7, 73–4, 87–8, 90–1, 94–6, 99–101, 106, 109–10, 118–20, 123, 125–7, 133–6, 138–9, 144, 146–7, 153–5, 192, 204–5, 220, 225, 229–30, 236, 239, 241, and government borrowing 82, 133–4, 144, 205, 208, 237, 244, and health 2, 9, 19, 21, 31, 35, 39, 58, 69, 104, 106–7, 138, 155, 162, 170, 201, 209, 230, 245, and honours 139, 158, 180, 185, 228, 234, 238, 242, and intercolonial conferences 78, 108, 114–15, 126, 129, 178, 191–2, 195, 212, and labour relations 17–26, 27–31, 34, 43, 53, 57–8, 63, 81, 85, 90, 99, 103, 107, 137, 172–3, 180, 200–1, 209–10, 242–3, 245, and marriage 2, 6–8, 44, 52, 54, 102, 110, 124–5, 128, 221, and oratory 134, 200–1, 215, 232, 241, and parliamentary elections 80–4, 115, 146–8, 165, 193, 219, 229–30, and pastoralism 115, 118, 126, 134, 143, 152, 163–5, 176, 187, 236, and Premiership 107, 109, 111, 117–18, 128, 133, 140, 143, 146–8, 150, 153, 178–81, 195, 198, 204, 207–8, 212–14, 218, 225, 236, 241–2, 244, 246, and Presidency of the Legislative Council 195, 216, 219, 222, 224, 226, 230–2, 237, 240–1, and probity 7–8, 93–6, 101, 105, 111, 119, 126, 130, 134–5, 146–8, 187, 203, 215, 225, 236–7, 243–4, and rail 45, 47, 72, 82, 85–6, 100, 116–17, 129, 133, 141–2, 145, 152, 158–9, 166, 169, 186, 194, 205, 208–9, 211, and religion 6, 25, 37–8, 44, 59, 64, 82, 88, 121, 143, 161, 165, 168, 171–2, 188–9, 205, 222, 237, 244, 245–6, and residences 9, 12, 26, 66, 69, 88, 92, 95, 97, 99, 121, 126–8, 155–6, 165, 172, 177, 203, 206, 209, 216–17, 219, 221–3, 228, 234–7, 239–40, and royal representatives 29, 84–6, 111–13, 115–18, 141, 143, 146–8, 151–2, 166, 179–81, 185, 187, 191, 195, 197, 207, 211–13, 216, 220, 222, 227–9, 237–8, 242, and Royalty 136, 139–42, 151–2, 160–1, 163, 166, 169, 179, 186, 211, 220, 227, 236, 240, and science 46, 50, 103, 105–7, 109, 122, 126, 129, 131, 145, 154, 164–5, 170, 175–7, 186, 188–9, 193, 204–5, 215, 245–6, and sport 69, 142, 151, 162, 206, and taxes and tariffs 108, 111, 115, 118, 129, 178, 195, 205, 208, 211, 237, and the Arts 69–70, 90, 102–3, 113–14, 128, 150, 157–60, 163–4, 171, 189, 195, 226–7, 229, 239, 245, and the environment 158, 167, 169–72, 178, 190, 210, and the Home visit 44, 52, 56, 67, 76, 83, 94, 101, 104, 106, 109–10, 113–14, 119–20, 127, 131, 136, 138–9, 146, 150, 152, Chapter 14, 174, 216, 243, 245, and the law 3, 6–7, 9, 12, 16, 19, 40–3, 64, 102, 109, 111–12, 124, 133–4, 186, 189–90, 208–9, 213–14, 222, 227, 230, 236, 240, and the Northern Territory 112–13, 117–18, 133–4, 141, 144, 146–7, 152, 210, 229, 241, 244, and the telegraph 96, 108, 112, 145, 165, 181–2, 184–6, 190, 194, 211, 227, 230, 241–2, and women 22, 69–70, 163, 166–7, 188–90, 206, 220, 228–9, 240–1, 245–6
Ayers, Henry (son of Harry Lockett) 176
Ayers, John Breaks 9, 12, 169, 239
Ayers, John Morphett 223
Ayers, Julian 194
Ayers, Lancelot Ernest 235
Ayers, Lucy Josephine (later Bagot, 'Josey') 73, 87, 157, 160, 170, 177, 212, 216, 239
Ayers, Lucy Lockett 2, 9, 69, 127
Ayers, Margaret Elizabeth (later Lungley, 'Maggie') 39, 87, 101–2, 115, 157–8, 160–1, 163, 168, 189, 198, 206, 223, 234–5, 239
Ayers, Mary Elizabeth (later Lady Lloyd) 152, 158, 223, 238, 252
Ayers Rock (Ulu<u>r</u>u) 1–2, 196–200, 222, 241
Ayers, Sidney Hurtle 223
Ayers, Sydney Breaks 87, 91, 122, 237
Ayers, William 2, 7, 9, 114, 119, 246
Ayers, William (Jnr.) 7, 9, 159

Bacon, Lady Charlotte 116
Baden-Powell, Robert (later Sir, then Baron) 164

Bagot, Charles Harvey, Captain 14, 85, 146, 152, 216–17
Bagot, Charles Hervey [sic], 235
Bagot, John (grandson of Captain) 216
Bagot, John Tuthill (nephew of Captain) 148–50, 198
Bagot, Lucy Josephine (née Ayers, 'Josey'), see Ayers, Lucy Josephine (later Bagot, 'Josey')
Bagot, Walter Hervey [sic], 223, 238, 252
Baker, Amy 234
Baker, Arthur 73
Baker, Arthur John 65, 73, 127, 189, 221, 234, 246
Baker, John 83, 134–5, 142, 146–8, 152, 176, 181–2, 189, 212
Baker, Lucy Lockett (later Rymill) 63, 102, 127, 156, 174, 221, 234
Baker, Margaretta Emily (née Lockett/Potts, 'Margaret') 6–7, 65, 234
Baker, Richard Chaffey (later Sir) 212–13, 219, 231, 236, 240, 242
Ballarat 64, 93, 115
Barrow, John 181, 187, 191–2
Beagle, HMS 4
Beck, Charles 25, 31, 34, 37, 46, 53, 104, 131
Beck, John 176, 177, 184, 201–3, 252
Bendigo 68
Beneficial Society's school, Portsmouth 3, 206, 245
Berry, Edwin 196
Bismarck, Otto Edward, Count von 128
Blyth, Arthur (later Sir) 108, 111, 114–15, 129, 150, 169, 179–82, 195, 198, 205, 210, 212, 234
Blyth, Neville 150, 176–7, 184, 252
Blyth, S. 3, 70
Bon Accord mine 89
Boothby, Benjamin, Justice 111–12, 133–4, 137
Boothby, Josiah B. 214
Boothby, William R. 214
Botanic Gardens, Adelaide 88, 190, 236
Boucaut, James Penn (later Sir) 179, 181, 186, 204–5, 208, 216, 244
Bray, John Cox (later Sir) 224, 227–8
Brougham and Vaux, Henry, Lord 3
Bruce, Talbot Baines 103
Brunel, Isambard Kingdom 129, 166
Bryant, Matthew, Captain 68, 91, 93, 121
Buckingham and Chandos, Duke of 28, 152
Buffalo, HMS 4, 206
Bullock, Frederick W. 253
Bullock, John 177, 194
Bunce Bros. 16, 33, 42, 45

Bundaleer sheep station 163, 165, 205
Burke and Wills expedition 112, 117
Burr, Thomas 21–3, 27, 30, 41, 47, 52, 89
Burra, 'Betsey', Miss 45, 47, 99, 103, 206, 227
Burra (or Burra Burra) Creek 14, 16, 18–20, 42, 52, 57–8, 60, 145
Burra Hotel 20, 29–30, 61, 63, 174, 209
Burra (or Burra Burra) mine or town 14–15, 17, 20–3, 26, 28–9, 31–5, 38–9, 41–2, 45, 47–51, 53–4, 56–8, 61–8, 72–3, 76–7, 80–1, 86, 88–93, 95–6, 99–100, 103–6, 108, 113, 116–17, 120–3, 126, 128–9, 131, 133, 138–42, 145–7, 149–50, 152, 154, 164, 172–5, 179–80, 182–3, 192–3, 200, 202–6, 209–12, 215, 217, 226–7, 230–1, 241, 244–6, 252

Californian gold rush 1849, 53, 58, 62
Cambridge 114, 138, 152, 164
Campbell, Alan, Dr 231, 239
Canada 7, 65, 114, 187, 235
Canowie sheep station 193
Cape Town 9, 40, 72
Carnarvon, Henry H.M. Herbert, The Earl 152, 165
Ceylon 21, 157
Challenger, HMS 4
Challoner, William 47, 57, 80, 129, 174, 177, 182, 184, 201
Chambers, James ('Jamie') 20, 25, 31, 64, 93, 100
Chambers, John 93
Chapman, William 177–8, 189, 220
Chartism 5–6, 31, 61, 84
Chile 36, 146, 176
China 32, 56, 205, 211
Chinese in Australia 103–4, 210
Chipman, Henry, Captain 21
Christchurch, North Adelaide 66, 93
Church, The Anglican 5–7, 25, 28, 30, 37, 59–60, 64, 66, 88, 93, 102, 140, 165, 170, 172, 207, 216, 235, 239, 245
Church, The Roman Catholic 143
Churcher, Elizabeth (née Lockett/Potts) 6–7, 114, 127, 234, 238
Churcher, John 6–7
Civil War, The American 103, 171
Clark, Caroline Emily (née Hill) 164, 189, 246
Clark, Caroline Emily 189, 246
Clark, John Howard 189, 213, 244, 246
Coke, James Charles 28, 34–5, 37, 39–40, 42, 44, 47–50, 52, 62–3, 65, 97, 183, 217, 222, 229
Coke, Jane 28, 34, 44

INDEX

Coke, (Margaret?), 'Mrs' 35, 40, 48–50, 75
Collegiate School of St Peter 22, 103, 128
Colonial Office (London) 10
Colton, John Blackler (later Sir) 208, 212–14, 242
Cowle, Amy Josephine (née Ayers) *see* Ayers, Amy Josephine (later Cowle)
Crimean War 70, 158, 171

Daly, Caroline, Lady 151, 161
Daly, Daniel 115
Daly, Dominick, Sir 111, 115, 127, 134, 141–3, 151, 237
Darlington, John 131, 145–6, 153–4, 160, 170, 174–7, 183, 192, 203, 215, 227, 230
Darwin, Charles 4, 165
Darwin, Port, (NT) 117, 144, 181, 186, 195
Davy, Edward, Dr 33, 35
Dickens, Charles 100, 108, 163, 195
Dilke, Charles, Sir 224
Dreyer, Georg Ludwig 18, 20, 32
Duff, John F., Captain 109–10, 153–4
Dutton, Francis Stacker 111, 118, 127, 144, 160, 212
Dutton, Frederick Hansborough 18

Edinburgh 72, 79, 167
Elder, Alexander Lang 144
Elder, Thomas (later Sir) 144, 150, 191, 204–6, 209, 216, 220, 235, 244
Elder, William 69
Elizabeth I 14
Elizabeth II 228, 251
Ellis, Elizabeth (née Ayers, 'Aunt Betsey') *see* Ayers, Elizabeth (later Ellis, 'Betsey')
Ellis, John 31, 43, 90, 165–6, 172
Elphick, William 46–7, 50, 80, 89
Emperor of China, barque 36, 39–40
English & Australian Copper Company, The (E&ACC, formerly the PCC) 31–4, 37, 42, 44–5, 47, 50–3, 56–8, 60, 62–3, 67–9, 72, 81, 105, 109, 131, 139, 183, 227, 245
English, James, Captain 36
Eugénie, Empress of the French 169
Ewbank, George 62, 67
Ey, Augustus, Captain 18

Falcon, HMS 116, 127–8
Federation of the Australian colonies 2, 78, 167, 192, 211, 229, 236, 238, 240, 251
Fenn, Charles 103
Fergusson, Edith, Lady (née Ramsay) 178, 222

Fergusson, James, Sir (Bart) 151–2, 158, 174, 178–82, 185, 191, 213, 222, 242
Fergusson, Olivia, Lady (née Richman) 222
Finniss, Boyle Travers 85, 112–13, 117, 144, 181, 241, 244
Fisher, Charles Brown (in partnership with brother James) 106, 109, 125, 163, 176
Fisher, George 96
Fisher, Hurtle 97
Fisher, James 106, 109, 125, 162–4, 176
Fisher, James Hurtle (later Sir) 8, 10, 12, 41, 71, 79, 83–4, 96–7, 106, 127–8, 180, 207, 240
France 3, 23, 31, 72, 91, 137, 142, 158, 165, 167, 169, 175, 211
Franco-Prussian War 165, 171, 175
Frankfurt am Main 67, 75, 91, 110, 154
Freemasonry 25, 27, 37–8, 61
Frome, Edward Charles, Captain 17
Furniss, Matthew Henry 182–4, 201–2, 215

Galatea, HMS 142, 151
Galvin, Bridget 223, 239
Gawler, George, Lt-Colonel 8, 10, 15
Gawler Town 85–6, 88–9
George III 133
George IV 3
George, of Wales, Prince (later George V) 220
Germany 18–19, 62, 67, 75, 110, 121, 128, 137, 142, 154, 160, 165, 167, 181
Giles, Ernest 196, 199–200
Gleeson, Edward Burton ('Paddy') 89
Glyde, Lavington 108, 180, 184, 195
Gosse, Agnes (later Hay) 200
Gosse, Agnes (née Hay) 200
Gosse, William, Dr 143, 162, 171, 199–200, 222
Gosse, William Christie 1, 196–200, 222
Gothenburg, steamer 207–8
government, local 10, 12, 17, 64, 83, 190, 210, 220–1
government, representative 12, 60–1, 64, 71, 73–4, 79, 83–4, 135, 183
government, responsible 2, 71, 77–8, 79–86, 88, 97, 100, 106–7, 111–12, 115–17, 127–8, 137, 145–50, 165, 180–1, 187, 189, 191, 193–5, 200–1, 205, 209, 212–14, 216, 219, 222, 224, 226, 228, 230–2, 234, 239, 204–41
Goyder, George 117, 144, 231, 236
Grace, W.G. 206, 226
Graham, Frederick <u>Malcolm</u> 72, 90, 121, 203–4, 211, 215
Graham, Henry Robert ('Harry') 49, 54, 90, 100, 203, 207, 211, 215, 224, 240

Graham, John Benjamin 20, 22–9, 31–40, 42–7, 49, 52–4, 56, 63, 65–8, 72–6, 82, 87–91, 93–100, 102–7, 109–12, 114, 120–2, 124–31, 133, 135–6, 138–40, 145–6, 150–6, 160, 169, 174–7, 182–4, 186, 192–3, 201–4, 206–8, 211, 215, 244
Graham, Louisa (née Rymill) 28, 44, 49, 53–4, 67, 71–5, 75, 87–90, 93, 97, 104, 106, 136, 146, 154, 160, 170, 175, 211
Graham, Louisa Maude 90, 160, 211, 215
Granville, George G. Leveson-Gower, The Earl 161
Great Exhibition 1851 49, 63
Grey, George, Captain (later Sir) 10–11, 14, 166
Grote, George, MP 5, 8
Gulf St Vincent 34, 42, 51, 141, 211, 215
Gwynne, Edward Castres 42, 51, 60, 74, 84, 112, 140, 209, 217
Gwynne, Edward Castres (Jnr.) 209

Hamlet (Shakespeare's) 114, 150
Hamley, Francis Gilbert, Lt-Colonel 146–8, 151, 166, 180
Handschuhsheim (Heidelberg) 110, 128
Hanson, Richard Davies (later Sir) 23, 51, 60, 140, 179, 208
Hare, Charles Simeon 116–17
Hart, John, Captain 179, 198
Hastings (Sussex) 154, 169, 216
Hay, Agnes (née Gosse) 200
Hay, Agnes (later Gosse) 200
Hay, Alexander 200, 219
Hayes, Catherine 70, 127, 163
Henderson Process, The 103, 105, 107, 109, 122, 126, 129, 215
Henderson, Robert 59
Henderson, William *see* Henderson Process, The
Hill, Rowland, Sir 164
Hill River sheep station 164, 176
Hindmarsh, John, Captain 4, 8, 10
Holder, Frederick W. (later Sir) 236
Holdfast Bay 9, 88, 110
Holy Trinity, North Terrace 38, 102
Hughes, Henry Kent 181, 183, 186
Hughes, Walter Watson, Captain (later Sir) 71, 102, 167, 191, 220, 244
Humble, S.W. 49, 97
Hutchison, Commander 113, 141
Huxley, Thomas Henry 165, 188, 246

Illustrated Adelaide Post 138, 144
Illustrated London News 23, 31
India 4, 32, 39, 45, 56, 205, 211, 217

Jacob, William 17

James, James 129, 155, 176, 238–9
Jervois, Lucy, Lady 228–9
Jervois, William F.D., Maj-General Sir 213, 215–16, 220, 227–9
Jewish community, The 128, 167, 171–2, 180, 188–9

Kapunda, mine and town 14, 85, 142, 208
Karkulto mine 54, 88, 91, 99
Kata Tjuṯa ('Mount Olga') 196, 200
Kaurna people, The 5, 10
Kent, Benjamin Arthur, Dr 70, 127, 171, 209
King, F.J. 227, 230
Kingston, Charles Cameron 205, 209, 236, 238, 240, 242
Kingston, George Strickland (later Sir, 'Paddy') 17–20, 31, 47, 59–61, 71, 80–1, 84, 152–3, 174–7, 180, 183, 191, 194–5, 202–3, 216–17
Kintore, Algernon H.T. Keith-Falconer, 9th Earl of 227–9
Kooringa township (Burra) and smelter 18, 24, 42, 44–5, 47, 58–61, 67–8, 105

Landseer, Edwin Henry, (later Sir) 28
Leach, Richardson & Company 33, 35–6, 48, 53
Leighton, Baldwin, Sir (Bart) 169, 189
Lesseps, Ferdinand-Marie, Vicomte de 166
Levi, Philip 128
Light, William, Lt-Colonel 8, 17, 83, 216–17, 225, 230, 245
Liverpool (UK) 21, 33, 89
Lloyd, Howard Watson (later Sir) 238, 252
Lloyd, Mary Elizabeth, (later Lady, née Ayers) *see* Ayers, Mary Elizabeth (later Lady Lloyd)
Lockett, William H. 7, 102
London 12, 16, 21, 22, 27–8, 31–4, 36, 44–5, 48–9, 53–4, 63, 65–6, 68, 72, 75, 78–9, 83, 90–1, 93, 95, 114, 118, 126–8, 133, 141, 144–6, 153, 159–61, 164, 166, 169–71, 175, 179–80, 182, 185, 188–90, 198, 201, 204, 206, 209–10, 212, 215, 223–7, 234, 238, 245–7, 251–2
Lungley, Arthur Robert 200, 206, 223, 226, 238
Lungley, Edith Amelia 209
Lungley, Margaret Elizabeth (née Ayers, 'Maggie') *see* Ayers, Margaret Elizabeth (later Lungley, 'Maggie')

MacDonald (sometimes Macdonald, and even McDonald), J.W. 24, 30, 71, 73–4, 93, 118, 243

INDEX   265

Married Women's Property Act 189–90, 217, 240
McKinlay, John 117
Melbourne 67, 70, 88, 92, 94–7, 102, 108, 111, 115, 123, 129, 131, 139–40, 190, 198, 210–11
Mildred, Henry Richard 60–1, 227
Mildred, Hiram Telemachus 227
Milne, Barbara (later Ayers) *see* Ayers, Barbara (née Milne)
Milne, William (later Sir) 181–2, 187, 195, 198, 210, 212, 215–16, 219, 234, 240, 252
Montez, Lola 69–70, 167
Moonta mine 100, 133, 186, 192
Morgan, William (later Sir) 178–9, 189, 204–5, 208, 212–13, 216, 218–19, 227, 237, 242, 244, 246
Morphett, Ada Fisher (later Ayers) *see* Ayers, Ada Fisher (née Morphett)
Morphett, Hurtle 252
Morphett, John (later Sir) 78–9, 83–4, 89, 97, 128, 137, 152, 178, 180, 195, 252
Morrison, James 144, 227
Murray, River 5, 57, 62, 97, 129, 142, 208
Musgrave, Anthony (later Sir) 195, 197, 211, 229
Musgrave, Jeanie (later Lady, née Dudley Field) 229
Mythology of the P.C.C. 51

Napier, Charles, General Sir 6
Napier Patent, The [James] 31, 33
New South Wales, Colony of 5, 50, 52, 115, 134, 144, 164, 192, 195, 211
New Zealand 103, 139, 166, 185, 191, 210, 217
Ngadjuri people, The 14
'Nobs' 15–17, 146
Nonconformity (religious) 3, 5, 26, 59–60, 141, 165, 171, 188–9, 213, 222, 244, 246
North (of England), The 4, 6, 89, 166–7
North, The Far (of South Australia) 93, 96, 100, 117, 134
North, The Mid (of South Australia) 24, 88, 104
Northern Territory of South Australia, The 112, 117, 133–4, 141, 146–8, 152, 181–2, 194, 205, 207, 210, 229, 241, 244

Old Colonists' Association, The South Australian 224, 240

Page, Evelyn Cameron (later Ayers), *see* Ayers, Evelyn Cameron (née Page)
Page, Thomas 160, 164, 166
Panama, Isthmus of 129

Paris 72, 158, 175, and its International Exposition of 1867, 91, 117, 142, and of 1878, 211
Parkes, Henry (later Sir) 232
Parkhurst Boys 15-16
Parkin, William 135, 140, 148, 189
Parliament, South Australian 2, 68, 71, 77–86, 88, 90, 97, 100, 106–8, 111–12, 115–18, 122, 128, 137, 139, 141, 144–50, 153, 179–81, 185–7, 189–95, 200, 202, 205, 208–10, 212–13, 215–20, 224, 226–8, 230–2, 234, 237, 239, 241–4, 250–1
Parliament, Victorian 71, 108, 114–15, 118, 129, 139, 210
Parliament, Westminster 5, 60, 71, 77–8, 85, 112, 160, 166, 179, 212, 230, 237, 250–1
Patent Copper Company, The (PCC) *see* English &Australian Copper Company
Patti, Adelina 163
Paxton, Edmund 163–4, 173
Paxton, William 30, 43, 47, 59, 61, 69, 95, 104, 127, 135, 159–60, 162–3, 167, 170, 172, 231, 245, 252
Peacock, Caleb 229
Peacock, William 60, 71, 90, 175, 177, 201, 202
Penny, Charles Mounsey 32, 42, 45, 51
Penny, Christopher Septimus 42, 45, 51
Petavel, W.H.A., The Reverend 209–10, 243
Phillips, Joseph 74, 92–4, 145
Pickett, Thomas 16
Pompurne mine 89, 91, 99
Portsea (Hants) 3–4, 8, 101, 114
Portsmouth 2–4, 6–7, 38, 70, 103, 114, 159, 163, 206, 227, 245
Potts, Anne (later Ayers), *see* Ayers, Anne (later Lady, née Potts)
Potts, Elizabeth (formerly Lockett, née ?) 5–9, 12
Potts, Frank 4, 9, 12–13, 17, 62, 127, 189, 221–2, 226, 235
Potts, Lawrence 5–9, 12, 17
Powell, William Llewellyn, Captain 35, 37, 44, 50, 62
Prince Alfred College 141, 221
Princess Royal mine 17, 19, 21, 45
Prospect House ('Graham's Castle') 24–5, 28–30, 36, 38–9, 43, 52, 66, 72, 93, 109, 177
Prussia 91, 117, 128, 165, 171, 175

Queensland, Colony of 104, 112, 210, 241

Ramindjeri people, The 5

Rangatira, steamer 114, 129, 151, 157
Redruth townships (Burra) 18, 58, 61
Reynolds, Thomas 116–18, 195, 205, 207
Richardson, barque 32, 36
Richman, John H. 9, 12–13, 55, 222
Roach, Henry, Captain 23–5, 29–30, 42–3, 46–7, 51–2, 54, 56–8, 60–1, 63, 68, 80, 89, 91, 100, 104, 121, 127–8, 131, 174
Robe, Frederick Holt, Lt-Colonel 59, 210–11
Roberts, Thomas, Captain 17–19
Robinson, William C.F., Sir 228
Rodda, Richard, Captain 107, 109
Russell, William Howard 158
Russo-Turkish War 213, 215
Rymill, Florence Edith 106, 120
Rymill, Frank 71–2, 74, 87, 95, 97, 102, 115–16, 118, 123–4, 126–7, 130–1, 136, 146, 152–3, 155, 193, 220–1, 227, 230, 235, 238, 252
Rymill, Henry, as Ayers admirer 54, 71–4, 76, 87, 89–95, 97, 99–102, 104, 106–7, 120–1, as Ayers antagonist 122–7, 129–30, 135–6, 138–41, 143, 145, 150, 152–6, 160, 174–7, 182–4, 186, 192–5, 200–4, 207–8, 211, 215, 221, 224, 227, 234–5, 238, 240, 243, 245
Rymill, John 97, 115
Rymill, Louisa (later Graham) *see* Graham, Louisa (née Rymill)
Rymill, Lucy Lockett (née Baker) *see* Baker, Lucy Lockett (later Rymill)
Rymill, Robert 74, 106, 115

Salisbury, Robert A.T.G. Cecil, 3rd Marquess of 64
Sanders, Robert, Captain 193, 201–4, 211
Sanders, William ('Mr Billy') 54–5, 104, 146, 189, 217, 246
Santo, Philip 116, 205, 209
Scarborough 166–7
Schneider, Henry William 31–2, 51, 63, 67–9
Schneider, Hortense 163
Schneider, John Henry Powell 31–2, 63, 67–8
Seafield Tower 206, 209, 219, 240
Short, Augustus, (Anglican) Bishop 25, 44, 50, 140, 206
Smart & Bayne 12, 16, 19, 29, 40, 253
Smith, Robert Barr 204–5, 225–6, 238
'Snobs' 15–17, 54
Sommer, Ferdinand von, Dr 19–20, 23
South Australia, Bank of 16, 54, 56, 67, 118, 144
South Australia, Commercial Bank of 225, 239

South Australia, Province of 2, 4–6, 8, 10, 12, 15–16, 21, 28, 31–2, 43, 45, 51, 59–60, 64–7, 71–2, 76, 78, 81–3, 91, 96, 104, 110, 112, 115–17, 122–6, 128, 133–6, 139, 151, 160–1, 166–7, 170, 173, 179, 192, 194, 200–1, 205, 210, 216, 224–6, 230, 232, 235–6, 240–1, 243, 245, 247, 251
South Australia, Savings Bank of 101, 108, 189, 226, 236, 239
South Australian Gas Company (SAGasCo) 108–9, 141, 157, 170, 216, 223–4, 229, 236, 238
South Australian Mining Association (SAMA) 15–23, 26–7, 29, 31–7, 39–50, 52–61, 63, 65–9, 71–6, 79, 83–4, 87–93, 95–8, 100–1, 103, 105, 107, 111, 118, 120, 122, 125–7, 129, 130, 139, 144–5, 150, 152–3, 155, 160, 165, 173–7, 180, 182–4, 189, 191, 193, 201, 203–4, 208, 210, 212, 217, 227, 229, 236, 238, 243, 245, 252–3
Special Survey 15–17, 58, 166
Spence, Catherine Helen 188–9, 240, 246
Spitty copper works 28, 31, 37, 51
'Squatters' 115, 117–18, 126, 134, 141, 187, 236
St John the Evangelist, Halifax St 30, 37–8, 64, 88
St Mary's, Alverstoke 7, 165
St Mary's, Kingston (Portsea) 7–8, 101
Stephens, Edward 16, 34
Stephens, John 41–3, 243
Stirling, John Lancelot (later Sir, 'Lance') 230, 238
Stocks, Samuel (Jnr.) 20–1, 43, 45–6, 52, 54, 204
Stow, Randolf Isham 116, 191, 208, 216
Strangways, Henry Bull Templar 117, 147–8, 152, 198
Streair, William 16
Streitman, Charles 214
Stuart, John McDouall 112, 225
Sturt, Charles, Captain 5, 35, 225
Suez 158, 166
Swansborough, William 154, 174, 177, 192–3
Swansea (Wales) 21–3, 27–8, 30–3, 35–6, 39, 42, 44–5, 48–9, 63, 67, 146, 166, 183, 217, 222, 229
Sydney 1, 42, 45, 55–6, 78, 108, 139, 143, 145, 192, 195, 211–12
Sydney Morning Herald, The 55–6

Telegraphy 96, 108, 112, 141, 165, 181–2, 184–6, 190, 193–4, 196, 199, 210–11, 224, 227, 230, 236, 241–2

Thomas, Andrew 50
Ticketings 28, 32–3, 36
Todd, Charles (later Sir) 141, 181, 185, 210, 227
Tolmer, Alexander 55, 67, 225
Tomkinson, Samuel 118, 155, 184, 202–4, 221, 224, 227, 231, 239
Torrens, River (Karrawirraparri) 10, 29, 51, 73, 137, 221
Torrens, Robert Richard 43
Trollope, Anthony 195

Uluṟu ('Ayers Rock') 1–2, 196–200, 222, 241
Unemployment 3–4, 8, 10, 129, 133, 137, 140, 145, 226
Unitarianism 165, 171, 188–9, 213, 222, 244, 246
United States of America 3, 5, 69, 103, 115–17, 157, 171, 187, 195, 211–12

Van Diemens Land (later Tasmania) 45
Victoria, Colony of 12, 62, 64–5, 67, 71, 93, 108, 112, 114–15, 118, 126, 129, 139, 141–2, 144, 194, 200, 210–11, 226, 228
Victoria, Queen 69, 71, 86, 91, 113, 137–9, 144, 161, 179, 186, 206, 226–8, 230, 236, 238, 240
Victoria Square, Adelaide 88, 137, 209, 216, 237, 242
Victory, HMS 2, 4, 103

Volunteers, The (militia) 102, 142

Wakefield, Edward Gibbon 4–5, 8, 51, 83
Wakefield, Port 51, 57
Wallaroo mine 100, 102–3, 214
Walters, Gregory Seale 31–4, 36–8, 41–2, 45–7, 50–1, 56–7, 62, 67, 105, 107, 118, 211
Ward, Ebenezer 231–2
Waterhouse, George Marsden 108
Waterhouse, John 25, 43, 53–4, 66, 237
Waterhouse, Thomas Greaves 54, 90, 145, 150, 237
Way, Samuel James (later Sir) 176, 208, 225–6, 232, 238
West, William 215, 238
Western Australia, Colony of 4, 10, 60, 78, 115, 140, 199
Westminster 5, 60, 71, 77–8, 85, 112, 160, 166, 179, 212, 230, 237, 250–1
Whitehall (British Government) 5, 8, 10, 15, 112
William IV 3, 5
Williams, Thomas Henry 37–8, 42, 47, 51, 62, 67, 105, 127, 166
Woolner, Thomas 171

Young, Augusta, Lady 29, 51
Young, Henry Fox, Sir 29, 43, 51
Younghusband, William (Jnr.) 48–9, 60, 71, 79, 83–4, 110, 128